AQUILA

HARRISBURG, ARKANSAS

FIRE-BOX BOILER WITH COLLAPSED FLUE TUBE.

Fig. 1.

Fig. 2.

Longitudinal Section

Horizontal Section

Fig. 3.

Fig. 4.

A

RUDIMENTARY TREATISE

ON

STEAM BOILERS:

THEIR

CONSTRUCTION AND PRACTICAL MANAGEMENT.

BY

ROBERT ARMSTRONG, C.E.

𝔉ourth 𝔈dition.

WITH AN APPENDIX OF RECENT AND IMPORTANT
DATA.

LONDON:

JOHN WEALE, 59, HIGH HOLBORN.

1862.

LONDON:
BRADBURY AND EVANS, PRINTERS, WHITEFRIARS.

PREFACE.

In order to render this book useful to those who employ, or are employed about, Steam Engines, I have included in it various rules and practical data formerly published in my "Essay on the Boilers of Steam Engines," and have endeavoured to embody the necessary information relative to such improvements in principle or construction as have come into practical use since the date of the former publication. This will, therefore, probably supersede the necessity of a new edition of that work.

R. A.

65, *Fenchurch Street,*
London.

CONTENTS.

CHAPTER I.

ON THE PROPORTIONS OF BOILERS SUITABLE TO DIFFERENT SITUATIONS
AND CIRCUMSTANCES, WITH EXAMPLES.

Page

Section 1. Introduction 1
2. Experimental Boiler 3
3. Steam Engine and Boiler Horse Power 4
4. Consumption of Fuel 6
5. The Flued Cylindrical Boiler 9
6. Area of Heating Surface. 16
7. The London or *Small* Cornish Boiler 18

CHAPTER II.

GREAT CHANGE IN, WITH REVIEW OF, THE PRACTICE IN THE
MANUFACTURING DISTRICTS.

Section 8. Wagon-shaped Boiler 23
9. Its Disadvantages and Disuse 27
10. On small Furnaces and excessive Firing . . . 29
11. General Mismanagement of Boilers, and Trade of Boiler
Making 32
12. On Enlarging the Furnace and Lengthening the Boiler . 36

CHAPTER III.

ON PROPORTIONING SURFACES, CAPACITIES, AND FIRE GRATES, WITH
RULES AND CALCULATIONS.

Section 13. Propriety of adopting a Square Yard of Surface for each
Boiler Horse Power 39
14. Proportioning the Fire Grate to the Heating Surface and
the Quality of the Fuel 41
15. Rules for finding the Horse Power, the Area of the Fire
Grate, and the Area of Effective Heating Surface useful
in altering or resetting Old Boilers . . . 46
16. On the Capacity of the Steam Chamber . . . 49
17. Causes and Effects of *Priming* 54
18. On Boiler Room generally . , ... 58
19. The Boulton and Watt Boiler 61
20. Capacity of Water Chamber 65

CHAPTER IV.

PRINCIPLES DETERMINING THE PROPER LENGTH OF BOILERS, WITH EXAMPLES, RULES, AND INSTRUCTIONS IN SETTING UP.

Page
Section 21. On great Extension of Surface and saving Waste Heat . 70
22. Great Extension in Length unnecessary. . . . 74
23. Principles governing the Length of Boilers . . . 75
24. Experimental Proofs in favour of Short Boilers . . 78
25. Proportioning Length to Quality of Coal . . . 81
26. Cornish and Butterly Boilers 83
27. General rules for Proportioning the Length of Boilers for stationary Engines 85
28. Boilers on the "Oven-Plan" liable to Explosion by producing surcharged Steam 88
29. Fire and Flame Bridges 93

CHAPTER V.

ON THE THICKNESS AND STRENGTH OF BOILERS FOR DURABILITY AND SAFETY, WITH EXAMPLES OF ACTUAL CASES.

Section 30. Practical limits to the Thickness of rivetted Boiler Plates 96
31. Rules for Proportioning the Strength to the Pressure and Stress on the Iron 99
32. Fairbairn's Patent Double-furnaced and Double-flued Boiler 102
33. Strength and Form of Internal Flue Tubes . . . 106
34. Marine Boilers 109
35. Galloway's Patent Double-furnaced *Tubular* Boiler . 111
36. Boilers at the Gutta-Percha Works 114

CHAPTER VI.

ON EXPLOSIONS, DEPOSIT OF SEDIMENT, AND INCRUSTATIONS.

Section 37. Explosions from Collapse of Flue 117
38. Collapse of Flue in a Low-pressure Fire-box Boiler . 122
39. Explosions from Incrustations 128
40. Deposit of Sediment 130
41. Calcareous Incrustations 135
APPENDIX 137

RUDIMENTARY TREATISE

STEAM BOILERS.

CHAPTER I.

On the Proportions of Boilers suitable to different Situations and Circumstances; with Examples.

SECTION 1.—INTRODUCTION.

WITHOUT agreeing in the opinion expressed by M. Pambour in the introduction to his excellent treatise on Railway Locomotive Engines, namely, that the theory of the steam *engine* itself has not yet been explained, we may with much truth affirm, that with respect to the steam *boiler*, even up to the present period, his words are peculiarly applicable, however successful has been the general practice.

Under such circumstances it is not surprising that a great variety of opinions are held on the subject. This difference of opinion relates not only to the form of boiler best adapted to supply the greatest quantity of steam with the least expenditure of fuel, but also to its dimensions or capacity suitable for an engine of a given number of horses' power; the mere arithmetic of the question remaining up to this day unsettled, or not generally agreed on : this latter subject we propose to consider principally in this and the two following chapters.

The only rule, if rule it be, for adjusting the dimensions of boilers amongst practical engineers and boiler makers, is to endeavour to have them larger than necessary ; hence it is a common observation with them, that a 10-horse engine should have a 12 or 15-horse boiler, or a 20-horse engine ought to have a 30-horse boiler, and so on. And we have known more than one extensive and successful manufacturer of engines, under 20-horse power, adopting the rule of making their boilers always 2-horse power more than the engines they are intended to drive. Certain vague notions have long existed amongst engineers, as well as with some writers on the subject, that there ought to be about 5 square feet of surface of water, or of the largest horizontal section of the boiler, for each horse power; and this mode of reckoning by water surface has more to do with the effective power of the boiler than at first view appears, but it only applies to the ordinary wagon boiler without internal flue. It has also been commonly considered that there ought to be about 25 cubic feet of space in the boiler for each horse power. The first of these data has become common amongst engine men as their only acknowledged rule for roughly estimating the power of a boiler, although both it and the mode of estimating by the cubic capacity have generally been scouted by scientific engineers; but with how little reason, while the latter have not furnished us with better methods, we shall see in the sequel.

After consulting Farey's, Tredgold's, and nearly all the other English works treating on the steam engine, besides discussing the matter with various eminent scientific persons and practical engineers, as well as with many of the most experienced manufacturers in Lancashire, we were convinced of the necessity of instituting and comparing a great number of experiments on a large as well, as on a small scale, and under a variety of different circumstances, in order, if possible, to deduce rules that might at least be applicable to those forms of boilers in general use.

Section 2.—Experimental Boiler.

Amongst our earlier experiments, on a small scale, was one
with a common furnace pot, or boiler, of cast iron, such as are
usually set in kitchens, sometimes called a *copper;* it was
capable of holding 18 to 20 gallons; the fire grate was
6 inches by 8, or one-third of a square foot in area, and the
whole of the heating surface exposed to the fire was about
3 square feet. Into this boiler was measured 2 cubic feet of
water, which was made to boil, after which it was found that,
by feeding the furnace with coal and the boiler with water,
and at the same time managing the draught of the chimney
so as to keep the water boiling nearly at a uniform rate, the
consumption of good coal was at the rate of $4\frac{1}{2}$ lbs. per hour,
and the quantity of water boiled away in that time was exactly
2 gallons, or very nearly one-third of a cubic foot; the
temperature of the water supplied, to make up for the evaporation, being 62°.

As it is very common to reckon the evaporation of a cubic
foot of water per hour as sufficient to furnish steam for one
horse power, we have only to multiply each of the foregoing
data by three to obtain the following proportions:—

1 cubic foot of water evaporated per hour requires

9 square feet, or 1 square yard of heating surface,

1 square foot of surface of fire grate, and

$13\frac{1}{2}$ lbs. of good coal.

We thus obtain what may be called a rough estimate of a
boiler of 1-horse power, and with an expenditure of fuel
approaching to what was formerly considered, by the oldest
disciples of the Boulton and Watt school, the proper, though
very ample allowance of 14 lbs. of coal per horse power per
hour.

This experiment, which was frequently repeated and verified
in a variety of ways, was of course made with an open-topped
boiler, which induced me to make several other experiments

with a variety of small closed boilers, more nearly assimilated to the condition of an ordinary steam boiler attached to an engine at work; the results were in all cases, whether under a pressure of 4 or 5 lbs. per square inch, or open to the pressure of the atmosphere alone, so nearly agreeing with the above, that, excepting increasing the quantity of water in the boiler, which increased the quantity of fuel used in getting up the steam, but at the same time lessened the other portion of the general consumption by lessening the difficulty of regulating the fire, the evaporation being steadier, there was no other alteration that materially affected the results as given above.

Section 3.—Steam Engine and Boiler Horse Power.

There is a remarkable coincidence in the figures expressing the dimensions of the apparatus and the effect produced in the above experiment, which renders it extremely convenient for forming a *unit measure* of steam boiler power; for we have only to conceive a vessel, say a cube for instance, of 3 feet square, or *a cubic yard* in capacity, with its lower side of *one square yard* in area, and exposed to the action of a fire on a grate of *one square foot* in area; let this vessel be nearly half filled with water, and then, after the steam is once got up, we have, by burning away 13 to 14 lbs. of coal in an hour, 1 cubic foot of water converted into steam, equal to the pressure of the atmosphere or a little above. This, if well applied, in a good Boulton and Watt engine, is well known to be at all times amply sufficient for one-horse power. It will do this even in a small short-stroke engine working without expansion, besides allowing various sources of waste and leakage from imperfect packing, &c.; but in larger and long-stroked engines, made to work to a certain extent expansively by means of " Lap " on the valves or otherwise, and not loaded much beyond the nominal horse power of the engine, as was the practice of the above celebrated firm for the last twenty years of its existence, the steam from a cubic foot of water usually produced an effect approaching more nearly to *one-and-a-half* horse power, thereby reducing the above rate of consumption below 10 lbs. instead of 14 lbs. per nominal horse power per hour. The entire mechanical force developed in the evaporation of a cubic foot of water is, however, much greater than this; but the above may be taken as the worst result obtained from the low pressure condensing engine throughout the cotton manufacturing district of Lancashire for the last quarter of a century, where the coal is generally of an inferior quality; and where, until within the last ten years, such engines, with steam not exceeding 6 lbs. or 7 lbs. per square inch above the atmosphere, and wagon boilers, were all but universal.

The measure of a steam engine horse's power, as originally settled by Mr. Watt, and to which both Farey, Tredgold, and all other engineers of eminence have agreed to refer as a standard, is a power equal to lift 150 lbs. 220 feet high per minute, or, what is equivalent thereto, 33,000 lbs. raised 1 foot high in the same time. It may be remembered that this was no fanciful standard of Mr. Watt's, but really taken by him as the average power exerted by a mill-horse travelling at the rate of 2½ miles an hour (or 220 feet a minute), and raising a weight of 150 lbs. by a rope, passing over a pulley, as appears by Mr. Watt's letter to Dr. Brewster, in the second volume of Dr. Robinson's Mechanical Philosophy, 1822. This standard horse power of Mr. Watt's ought not to be departed from, it having the merit of great simplicity and convenience. *One hundred and fifty* pounds being exactly divisible by *six*, which is the effective pressure per *circular inch* on the piston (nearly equal to 8 lbs. per square inch) considered most suitable for the best modern engines working with little expansion, gives exactly 25, which is the number of circular inches per horse power in Boulton and Watt's original 40-horse engines very nearly. Hence, a cylinder of 5 inches diameter ($= 25$ circular inches area), with an effective pressure of 6 lbs. per circular inch on a piston travelling at the rate of 220 feet a minute, will give out one-horse power.

Section 4.—Consumption of Fuel at Factory Engines.

Regarding the quantity of fuel ordinarily consumed by steam engines in the factories of Yorkshire and Lancashire, an immense deal of misconception and ignorance has hitherto prevailed in some quarters, which it will be worth some trouble to remove. Considering the great importance of correct statistics on this branch of the subject, I took some pains in endeavouring to arrive at the truth previous to making the statements on this head in my first practical Essay on Boilers, published more than twelve years ago. In obtaining the best authorities in Manchester there was no difficulty, for in the hundred cotton factories of that city I had direct access to the coal account of nearly one-half, besides having been several years previously well acquainted with the minutiæ of the working economy of their engines and boilers. Besides this, I had a similar acquaintanceship with a still greater proportion of the factories in Preston, Bolton, and Stockport, as well as throughout the cotton manufacturing district generally. Moreover, in stating that the consumption of fuel per nominal horse power of the engines in the cotton factories, was something *under* 10 *lbs. per hour* on the average, I did not leave the fact to rest on assertion and authority alone, however indisputable, but gave various reasons for believing that it could not be more than 10 lbs.

Notwithstanding this, some authors of large books, with no mean pretensions to accuracy, as well as writers in the public scientific journals, have gone on from that day to this reiterating the extravagant assertion, that the "*ordinary*" consumption of coal in the manufacturing districts is 15 lbs. to 18 lbs. per horse per hour, while the same quantity of work is performed by the Cornish, French, and Belgian engines, for $2\frac{1}{2}$ lbs. to 3 lbs., or apparently *one-sixth of the Lancashire consumption*. Now, to say nothing of Belgium or France, the engineers of the latter country, being supplied with the same tools, the same means, and having generally speaking more science, and some other advantages, *ought* really to go a step in advance of the native country of the steam engine. The great disparity between the Cornwall and, Lancashire statements might have been reconciled by any common-sense observer asking the simple question, Are not the Cornish engines of very great power, and generally not half loaded, while the Lancashire engines are frequently loaded to double their nominal power?—which would at once account for *two-thirds* of the anomaly, while the rest is amply accounted for by the great superiority of the Welsh coal used in Cornwall over the "Slack" and "Burgey" used in Lancashire. Although not very precise, this is a true and very nearly correct statement of the case, as we shall have several occasions to prove in the course of this work.

It ought also to be borne in mind that the 14 lbs., or $\frac{1}{8}$th of a cwt. of coal per horse per hour spoken of, is the whole or "gross" consumption, which includes all that· used for getting up the steam every morning, as well as to supply steam sufficient for heating the building. This, in the practice of the best-regulated cotton-mills in Lancashire, is found to be very considerable, even during summer, in addition to steam used for a great many other heating purposes peculiar to the cotton manufacture. One-eighth of a cwt. per hour is equal to 10 cwt., or half a ton, for 80 hours ; and as 13$\frac{1}{2}$ hours per day (when the engine works 12 hours), or rather more than 80 hours per week, is about the length of time that the boilers are at work during each week, in cotton mills, we may state generally that half a ton per horse power per week is the average gross consumption.

Now, in order to obtain the average *net* consumption of the cotton-mill engines in Lancashire, we must first deduct the proportion required for "steaming" the mill, which, as it depends on a great many varying circumstances, is not easy to estimate. It is generally supposed that in the best constructed modern factories it is impossible to be less than

from 10 to 15 per cent. There is reason to believe from accounts of coals kept where separate boilers are used for this
purpose, as well as from measurements made of the quantity
of condensed steam-water obtained from the steam-pipes,
that the average quantity of fuel required for heating a cotton mill cannot be so little as 20 per cent. of the whole consumption, and in one instance I found from the last-mentioned
mode of estimating, that it was near 30 per cent. ; but, to be
on the safe side, we will only take it at 15 per cent., when it
will leave

$14 - (\frac{15}{100}$ of $14 =) 2 \cdot 1 = 11 \cdot 9$ lbs. per horse power per hour
for the consumption on account of the engine alone.

When the above estimates were made, the engines in cotton
factories uniformly worked 69 hours per week ; which calculation it will be best for our present purpose to adhere to, since
the Short Time Act has introduced a very uncertain element,
that is, the want of uniformity in different mills ; the only
thing certain at present ascertained, is, that the saving in fuel
is nothing like equal to the diminution of the time. This
reckoning allows only 11 hours during the week for getting up
the steam every morning, and for stoppages at meal times,
which gives one hour and three-quarters per day, with half-an-
hour extra for Monday morning. All persons practically conversant with this subject, know this is a very low estimate,
and would not reckon two hours sufficient for this purpose, in
an average of mills, and taking the year through. Considering
also that there must be at least as much coal laid on per hour,
in getting up the steam, as is used per hour after, we are
certain to be within the truth by allowing 15 per cent. of the
whole consumption for that purpose. Therefore, we have
$11 \cdot 9$ less by $2 \cdot 1 = 9 \cdot 8$ lbs., or say nearly 10 lb. per horse per
hour, as the net consumption of coal while the engine is at
work. There is great difficulty in coming at the exact proportion in cotton factories, on account of the consumption for
this purpose being unavoidably mixed up with that for steaming the mill before mentioned, which is always the greatest in

the morning at the same time when the steam is being got up
for the engine; consequently very great discrepancies of
opinion are met with on the subject.

Whatever may be the exact amount of each of the two
items we have been treating of, namely, for getting up the
steam and for heating the mill, we are well assured that the
combined amount of the two, 30 per cent., at least, ought to
be deducted from the gross amount of the coal used, in order
to get at the net consumption while the engine is at work,
leaving it, as already stated, under 10 lbs. per horse power
per hour.

Now, it must not be forgotten that all the above statements
and estimates are founded on the *nominal* horse power of the
engines only, while it is well known that the cotton-mill
engines are very commonly working to double their nominal
power; and usually considered light loaded if only working to
one-half more, which latter proportion I would say is within
the average load on the engines throughout the cotton-mill
district. This consideration at once reduces the apparent
consumption by one-third, or from 10 lbs. to between 6 and
7 lbs. per horse per hour.

SECTION 5.—THE FLUED CYLINDRICAL BOILER.

The second experimental boiler I took some pains in in-
vestigating was of a cylindrical shape, a kind commonly used
to high-pressure engines, as this one was, and generally considered very safe and strong. Its dimensions were as follow
(see Fig. 1). The shell of the boiler was 5 feet in diameter
outside, and 9 feet long; it contained a cylindrical flue tube,
running through the lower part from end to end, of 18 inches
inside diameter. The fire grate was 3 feet 6 inches square,
or 12¼ square feet area, and placed under the boiler close to
one end. The flame or smoke, after passing from the fire and
under the boiler bottom to the further end, rises up and returns through the inside flue to the front end above the fire-
door, where it is allowed to divide itself into two brick flues—

B 5

Fig. 1.

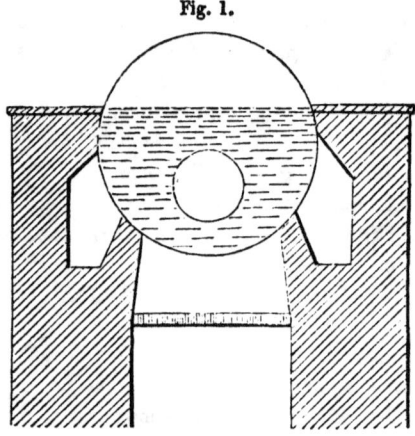

one on each side—through which it is conducted along the sides of the boiler again to the back; here the two currents again unite, and proceed in one main flue to the chimney. When the current of smoke and hot air is thus divided into two flues, it is called a "*split* draft," and when it is continued round in the same direction, it is called a "*wheel* draft."

This boiler, according to the common horizontal measurement of 5 square feet, would be equal to *nine*-horse power; but engineers who are in the practice of making this kind of flued boilers usually consider the diameter of the inside flue tube as equal to so much added to the width of the boiler, a mode of estimating its evaporating power which gives a result nearly correct. For in cases like this, where the boiler is short in proportion to its length, a great portion of the flame must act as effectually against the top of the inside flue as it would have done against a continuation of the boiler's bottom. So near, in fact, to the truth has all examinations of the power of boilers by this rule proved to be, that with a proper divisor and due allowance being made for the capacity of the boiler as respects room for water and steam, I do not hesitate to recommend its general adoption for boilers of this description. With respect to the proper divisor, it may be remarked, that although many practical engine and boiler makers have generally used the number 5, which gives, as they term it, 5 square feet of "*surface of water*" per horse power, others, and amongst them the late Mr. Benjamin Hick, of Bolton, were accustomed to give 5½ square feet, both in cylindrical and wagon boilers; while many country manufacturers, when ordering new boilers, have insisted on having 6 feet of water surface per horse; remarking, that after following that rule for many years they never got a boiler too large.

We shall also very soon show that 5¾ square feet of "water surface" per horse is perhaps more eligible as being applicable to a greater variety of boilers.

According to Mr. Hick's rule, the power of this boiler will be calculated as follows :—

$$
\begin{array}{lr}
\text{Width or diameter of boiler} \quad . \quad . & 5 \ \ \text{feet.} \\
\text{Ditto of inside flue} \ . \ . \ . \ . \ . & 1\tfrac{1}{2} \\
\hline
& 6\cdot5 \\
\text{Multiply by the length} \quad . \ . & 9 \\
\hline
\text{Divide by } 5\tfrac{1}{2} \text{ or} \quad . \ . \ . & 5\cdot5)58\cdot5(10\cdot6 \\
& 55 \\
\hline
& 350 \\
& 330 \\
\hline
\end{array}
$$

This gives rather more than 10½ horse power.

The evaporation in this boiler was ascertained to be very regularly from 10 to 11 cubic feet of water per hour, with from 12 to 15 lbs. of coal for each cubic foot ; say 4½ to 5½ lbs. of water evaporated to each pound of coal consumed. It was attached to a high pressure or non-condensing engine nominally of 12-horse power, but certainly when at the best not doing so much as 10-horse power, although working with a pressure in the boiler of above 30 lbs. per square inch. One cause of the great want of economy in this engine arose from what used to be described, before the times of metallic pistons and railways, the "*natural* defects" of the high-pressure engine, namely bad "packing" and bad "exhaust," by which nearly double the quantity of steam was used that would have been required in a low-pressure engine to do the same work, besides requiring very close attention, and what is called "*hard* firing," to get the proper speed out of the engine. This "hard firing," which for non-professionals may be translated to mean "hard work at firing and stoking," again superinduced other evils, of which by far the greatest was the "*priming*" of the cylinder with water, and sometimes *dirty* water too, instead of steam, of which and the cause of it more in the sequel.

As this work is meant to be practically useful to the working mechanic and engineer, we make no apology in giving the following direction and formula for finding the power of any boiler of this kind on the common carpenter's slide rule, in addition to the arithmetical operation already given.

Set the divisor or gauge point G P, for the power = $5\frac{3}{4}$ upon A, against the length L = 9 upon B; then against the diameters of the boiler and flue, added together, $d = 6\frac{1}{2}$ upon A, is the horse power H P upon B = 10·2.

A	G P = $5\frac{3}{4}$	$d = 6\frac{1}{2}$
B	L = 9	H.P. = 10·2

As there is no necessary calculation about a steam engine that cannot be done with a few simple operations of the slide rule, every workman's attention ought to be urged to it for many reasons, but mainly because it gives him ready facilities in comparing the proportions of engines and boilers by different makers without waiting for long laborious calculations, only to be done *after* some of the main circumstances affecting the action of the engine are forgotten. For this purpose no intelligent engine driver or stoker ought to be without a slide rule, and for that purpose there is no absolute need of carrying a "two foot;" one sufficient for all common purposes is only 4 inches long, with a reversed slide, and may be carried in the waistcoat-pocket. It has both a direct and a reversed line on the slider, which gives it several advantages, besides that of going into one-half the compass; and, as we shall make some use of it in this work, we shall repeat the operation as an example.

Reverse the slide, then set the diameter d of the boiler and flue upon A, against the length L, upon ɔ, then against Mr. Hick's divisor or gauge point H $= 5\frac{1}{2}$ upon A, is the horse power H P upon ɔ.

A	H $= 5\frac{1}{2}$	$d = 6\frac{1}{2}$
ɔ	H P $= 10\frac{1}{2}$	L $= 9$

In the same manner, without altering the slide, you have opposite the other three divisors their corresponding results, the same as if set down in a table thus :—

The divisor 5 gives the result . . $11\frac{3}{4}$ horse power.
   Ditto $5\frac{1}{2}$ ditto . . . $10\frac{1}{2}$ ditto.
   Ditto $5\frac{3}{4}$ ditto . . . $10\cdot2$ ditto.
   Ditto   6   ditto . . . $9\frac{3}{4}$ ditto.

Another advantage of the reversed slide is, that either the divisor or the result may be taken on either line A or ɔ, indifferently, without the risk of any mistake.

We shall now compare the dimensions of this boiler with the unit measure derived from the first experimental boiler.

	Circular feet.
Diameter of boiler squared, or 5^2 . . .	$= 25$
Deduct area of end of flue, or $(1\frac{1}{2})^2$ . .	$= 2\frac{1}{4}$
Net area of boiler end	$= 22\frac{3}{4}$

This multiplied by ·7854 . . . . $= 17\cdot86$ sq. feet.
Again multiplied by the length, 9 . $= 160\cdot74$ cub. feet.
Which, divided by 27 . . . . . $= 5\cdot95$ cub. yards.

The last two lines give the total capacity of the boiler for water and steam nearly *six* cubic yards. This, equally divided between steam and water, gives $160 \div 2 \div 10 = 8$ cubic feet of water per horse. But, as in the first experimental boiler, it was found by several trials that a greater proportion of water was more economical, as tending to balance any want of uniformity in feeding the fire, or in the action of the force pump which supplied the boiler with water, and on that account the quantity of water really worked with was about 4 cubic yards, or nearly 11 cubic feet, per horse power.

Since the advent of locomotives and tubular boilers, many engineers will object to this quantity of water as unnecessarily large. The results stated, however, were from careful experiments made purposely to see what quantity of water was the best under *common circumstances*, as in this boiler, which was not supplied with water by self-acting feed apparatus, and therefore not constantly uniform; had that been the case, and had the fire been also supplied in a similar manner with fuel, there is every reason to believe the quantity of water might have been greatly reduced with advantage.

The water room being thus increased, the steam room was of course diminished to about one-third of the capacity of the boiler; a space found to be quite small enough even when the engine had only half its load on. This evil was in some measure remedied by a method of working, which partially prevented the occasional boiling over or " *priming over* " of the water from the boiler into the cylinder, although accompanied with a considerable deduction from the power of the engine, by a process which is technically called " *wire drawing the steam*," or, in other words, "throttling" or partially closing the passage for the steam through the main steam valve or nozzle in the steam induction pipe leading from the boiler to the engine; one of the effects being that in order to get the same work out of the engine the steam required to be worked a great deal higher in the boiler than in the cylinder. This wire-drawing process was at one time much lauded by some advocates of high-pressure engines and steam-carriage projectors as an important discovery; it is simply, however, only an expedient to enable an engine to be worked with too small a boiler; and if not with an increased consumption of fuel, certainly with increased tear and wear as well as danger of bursting the boiler.

The above remarks on wire-drawing steam are nearly in the same words as a similar statement in the first edition of the "Practical Essay;"—that statement, however, like many others in that work, to be properly understood, requires some modification and explanation which greater consideration and more correct observation of results obtained by the application of the steam-engine indicator now enables me to give. It has been pretty well known to engineers for the last thirty years, that great economy of steam and consequently of fuel was to be obtained by working steam expansively, that is, by causing the steam to be shut off some time before the piston reached the end of the stroke, the remainder of the stroke being accomplished by the expansion of the steam already within the cylinder. It is, however, not so well known—at least not so universally admitted—that all engines working with a crank and flywheel, if working at a good speed and with a moderate load, have at *all times* worked the steam expansively, and must necessarily do so, even without any lap on the valves or other arrangement for cutting off the steam. This arises from the variable velocity of the piston, which at the beginning of the stroke is very slow; but as the motion becomes quicker towards the middle of the stroke, the aperture for the admission of the steam becomes too small to keep up the pressure, and the steam in consequence expands. The continued admission of the steam, however, as the piston arrives at the slow part of its motion again, and the crank approaches the centre, would cause a greater expenditure of steam without a corresponding advantage. Engines, therefore, have usually been made with a little lap on the steam side of the slide valve, in order to obviate this defect. This is made very plain by examining the indicator diagram of a throttled engine working with a moderate lap on the slide, and its action will be found to be very nearly the same as that of an engine working with a considerable amount of expansion obtained by a separate expansion valve.

So much for economising steam by working with a throttled engine; but it will be well to look to the disadvantages thereby occasioned with respect to the evaporative power of the boilers, to say nothing of the danger; for it is quite certain that the risk of bursting the boiler increases directly with the pressure. Other things remaining the same, the boiler is also weaker as its diameter is increased; although giving a large capacity of steam chamber need not necessarily increase the diameter of the boiler; for, as we shall soon have occasion to show, a separate cylindrical steam chamber is in every respect more eligible. The main reason, however, why the particular way of working the engine affects the evaporative economy of the boiler is this: the more the steam is throttled, the slower the water boils; and, other circumstances being the same, it is quite evident that the less rapidly ebullition is carried on the less steam is produced, and the more coal is wasted—an unanswerable argument against the advocacy of great, rather than good boilers, which we shall have further occasion to refer to.

SECTION 6.—AREA OF HEATING SURFACE.

The most important consideration that affects the calculation of the evaporating power of the boiler is the quantity of *heating surface*, or surface exposed to the hot air, and the proper method of measuring it. The usual method of estimating the *whole area* of surface exposed to the hot air as heating surface, has been a fruitful source of mistakes and misapprehensions among steam engineers: because, as heat is with difficulty made to pass downwards to any useful purpose, it is plain that such portions of the surface only as are exposed to the *upward* action of the flame and hot air, ought to be considered as *effective heating surface* to its full extent. With regard to the upright or perfectly flat vertical portions of a boiler there may be a question, but it seems to be agreed to by all who have attended to this subject, that at any rate the ordinary side surface is not more than one-half as effective in generating steam as the *under* surface, and we have assumed this to be the ratio in the following calculation, which is inserted merely as an example of a method of computing the heating surface, which is found to be sufficiently correct for practice, when applied to the ordinary forms of boilers in general use. In the boiler under consideration—

	Sq. ft.
The area of the boiler bottom $9 \times 3\frac{1}{2}$	$= 31\cdot5$
The two sides together $1\frac{1}{2} \times 18 = 27$, take half .	$= 13\cdot5$
The two ends about 12 . . . . . ditto .	$= 6\cdot0$
The inside flue $1\frac{1}{2} \times 3\cdot1416 \times 9 = 42\cdot4$; of this only the upper half can be considered as actual or effective heating surface	$= 21\cdot2$
	$72\cdot2$

This sum divided by 9 gives 8 square yards as the total effective heating surface of the boiler, which is considerably less than 1 square yard for each horse power, taking the boiler to be $10\frac{1}{2}$ horse, according to Mr. Hick's rule. This in part accounts for its extravagant consumption of nearly 15 lbs. of coal per cubic foot of water evaporated, being $1\frac{1}{2}$ lb. more than was required in our first experimental boiler, although the coal was of the same quality in both cases.

We have, however, not yet taken into account the most important element, if not of the *economy* at least of the *power* of the boiler, and that is the size of the fire grate, which, being $12\frac{1}{4}$ square feet in area, gives more than 1 square foot for each horse power. Hence, by the area of effective heating surface, the boiler is a little more than 8 horse, whilst according to the area of fire grate it is above 12; therefore, we may take the mean, say about 10-horse power, which gives about the same proportions as the data assumed from the first experimental boiler.

It may here be remarked that, had the situation in which it

was placed admitted, this boiler, for a 12-horse engine, ought to have been made 12 or 14 feet long, and a little less than 5 in diameter, and then it would have done as well without as with the flue-tube; excepting that the latter makes a good stay for the boiler ends.   The space fixed on for the reception of the boiler, however, admitted of no extension in any direction.   Nothing is more difficult to arrange economically than the packing of an efficient and durable boiler into too limited a space.   The foregoing is an example of one belonging to a small work (not a cotton factory) in the centre of the city of Manchester; and is, perhaps, the best arrangement that could be made with the means at command, and where the price of good coals is usually about 6s. per ton; the large consumption of coal being the principal evil entailed by the difficulties of the case.   In some places, London for instance, where coal is three times the above price, a different arrangement is more adviseable; and it may be interesting in a practical point of view to know not only what ought to be done in a similar case under such altered circumstances, but (as I happen to have an opportunity) also to show what actually has been recently done by one of the principal firms of mechanical engineers in the metropolis.

SECTION 7.—THE LONDON OR *SMALL* CORNISH BOILER.

This boiler is described in detail, and the various data connected with it given for the purpose of adverting to hereafter. It is called the "London" boiler for distinction's sake only; because the space in which it is placed is so very nearly (except in height) of the same extent as that occupied by the Manchester boiler described in the last section, and because I consider it very judiciously arranged for meeting the difficulties of a confined situation, such as frequently occur in the densely crowded localities of large cities.   It is on that account also a useful practical lesson for boiler engineers.

The shell of this boiler is $5\frac{1}{4}$ feet in diameter, and $9\frac{1}{2}$ feet long, being 3 inches wider, and 6 inches longer than the

Fig. 2.

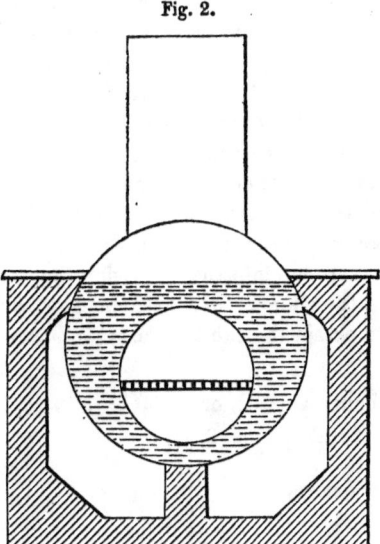

Manchester boiler just referred to. It has a cylindrical fire-tube flue running through it of 2 feet 10 inches diameter; which, like all other Cornish boilers, contains the furnace and fire-bars. These dimensions, according to Mr. Hick's rule, (p. 13), and the following operation by the slide rule, give

A	G. P.	$5\frac{1}{2}$	$d = 5\cdot25 + 2\cdot8 = 8$
Ɔ	H. P.	14	L $= 9\frac{1}{2}$

the nominal power of the boiler as equal to fourteen horses, or one-third more than the Manchester boiler, so far as the heating surface is concerned.

The fire grate, consisting of 25 fire bars of $\frac{7}{8}$th inch thick on the top face, with $\frac{3}{8}$th inch draft spaces between each, is 4 feet long by 2 feet 9 inches wide, or 11 square feet in area, and it is so arranged with a removable brick bridge that the whole or any less area of grate can be used that is found most economical, and is at present working with $3\frac{1}{4}$ feet length of fire-bar, or 9 square feet in area. The greatest rate of com-

bustion when the draft is full on is about 16 lbs. of the best
Newcastle coal per square foot per hour; the draft into the
chimney flue, which has no connection with any other fires,
being at the time equal to the pressure of a column of water of
0·46 inch high, and the actual velocity of the smoke up the
shaft about 20 feet per second.

The current of flame and hot air in this boiler first passes
from the fire through the flue tube to the back end of the
boiler, whence it returns under one side of the shell to the
front, where it crosses into the other side flue, through which
it proceeds to the chimney in a continuous or wheel draft.

In order to obtain the cubic capacity we have—

		Cir. ft.
Diameter of boiler squared, or $\overline{5\cdot25}|^2$	=	27.56
Deduct area of end of tube, $\overline{2\cdot83}|^2$	=	8
Net area of boiler end		19·56

This multiplied by ·7854 = 15·36 square feet, and again
by 9½ feet = 145·9 cubic feet for the capacity of the barrel of
the boiler. But there is also a steam dome or box from which
the steam pipe proceeds, erected on the top of the boiler, the
ostensible purpose of which is to prevent priming.

This steam dome is oval, in section 2 feet 6 inches by 3 feet
2 inches and 3 feet 10½ inches high, equal to about 24 cubic
feet, and making the total capacity of the boiler nearly 170
cubic feet, which is only about 10 feet more than the Manchester boiler, and consequently, like that, it is found to be
very liable to prime. It appears also that allowing 6 inches
depth of water over the top of the flue—which is the least that
ever ought to be allowed for safety—the quantity of water
worked with is equal to about half the net capacity of the
barrel of the boiler, or 73 cubic feet; and calling the boiler
14-horse, it gives only 5·2 cubic feet per horse power, or only
about half that required by the Manchester boiler, which is a
remarkable difference. On the other hand, the steam-room is

larger, being, including the dome, not less than 7 cubic feet per horse power.

If the heating surface of this boiler be calculated in the same way as before, in order the better to compare it with the Manchester boiler, and taking the inside tube flue first—we have $2\cdot8 \times 3\cdot1416 \times 9\frac{1}{2} = 83\cdot55$ square feet; but taking the upper half only as effective heating surface, as before, it is $= 41.77$.

In measuring the side surface in all boilers set up in this way, there is a difficulty in fixing the proper line where the bottom surface ends and the side surface begins, but, as the side flues are gathered in about 6 inches above the central line of the boiler, it appears fair to consider the side to extend to the same distance below, as that would make it equivalent to so much vertical surface.

The sides, therefore, will be 1 foot deep by $2 \times 9\frac{1}{2}$ long, half of which being effective $= 9\cdot5$ square feet.

The bottom surface will of course consist of all the rest of the shell exposed to the heat, except that portion occupied by the central supporting wall; or
$5\cdot25 \times 3\cdot1416 \div 2, - 1\cdot75 \times 9\frac{1}{2} = 61\cdot718$ square feet, which, added to the tube and side surface, gives the total effective heating surface $= 112\cdot988$. This divided by 9 gives $12\cdot55$ square yards.

In comparing this calculation with that of the Manchester boiler (at page 17), we begin to see the reason of the great superiority of the London one, not only in having a larger proportion of heating surface, but in having a greater portion of the lower part of the shell *effective*, principally arising from the absence of the two supporting walls, which in the former case confine the bottom heating surface to a comparatively narrow strip, less, in fact, than is obtained by the top of the flue only of the London boiler.

The general results of the above comparison show that the London boiler, which is so very little larger, and, with the accompanying brickwork, really goes into less space, has 50 per cent. more in area of effective heating surface, to little more than half the quantity of water, than the Manchester boiler. The economical results in practice are of course found to be correspondingly great. The quantity of water evaporated in the London boiler being considerably greater for each pound of coal consumed,—in fact nearly double, or equivalent to saving one-half the coal in doing an equal quantity of work. It is true that a difference in the quality of the coal used in each case no doubt contributed its share in producing the different results, although in both cases the best coals obtainable in their respective districts were necessarily used. There were also some instructive points of difference connected with the furnace and draft, to be adverted to hereafter. At present we shall only remark that, in estimating the heating surface, Mr. Hick's rule of $5\frac{1}{2}$ square feet of water surface or of horizontal section per horse power, which is quite suitable to the Manchester boilers, that admit of using a greater quantity of cheap coal on a larger fire grate *underneath* the boiler, does not appear so applicable to the London boiler, with its fire grate of more limited dimensions *withinside* the boiler, and on that account necessarily requiring a better quality of fuel. This difference of construction, as respects the furnace requiring a better quality of coal than is obtainable in Manchester, again creates the necessity of a larger area of heating surface, in order that the smoke or other products of combustion may pass off into the chimney at the same moderate temperature in each case. On this account mainly, there can be little doubt that *six* square feet of section per horse power would be quite as eligible for one description of boiler as *five and a half* is for the other. Thus calculated, the nominal power of the two boilers will stand thus :—

Manchester . . . $\dfrac{5 + 1\cdot5 \times 9}{5\cdot5}$ $= 10\cdot2$ horse power.

London. . . . $\dfrac{5\cdot25 + 2\cdot8 \times 9\cdot5}{6} = 12\cdot7$ ditto.

In recommending, several years ago, an intermediate number as a divisor that will be more generally applicable to all kinds of stationary boilers,—namely *five and three-quarters*, or, what is preferable, perhaps, 5·73 square feet, which gives exactly one square yard of effective heating surface per nominal horse power,—I did not then anticipate so strong a corroboration of its propriety as the above two cases exhibit. I shall therefore have the less hesitation in using it in this work as the "unit of steam boiler power,"—as well as to urge its general adoption by English engineers.

CHAPTER II.

Great Change in, with Review of, the Practice of the Manufacturing Districts.

SECTION 8.—WAGON-SHAPED BOILER.

IN order to examine more accurately the proportions that have been generally adopted in practice, in connection with the common-law pressure condensing engine, it will be best to take an instance of a plain wagon-shaped boiler without any inside flue (see Fig. 3), say 20 feet long, 5 feet wide, and 6 feet 8 inches deep. This description of boiler has been until recently in very general use throughout the manufacturing districts of Yorkshire and Lancashire, and with these dimensions it is called a 20-horse boiler, being well known to be fully equal to supplying steam for a 20-horse engine, with a very moderate consumption of fuel. They are also commonly made from 16 to 24 feet long, with the same depth and width, being considered as many horses' power as feet in length.

In the furnace of this 20-horse boiler the fire grate is 5 feet long and 4 feet wide, containing a set of fire bars in a

Fig. 3.

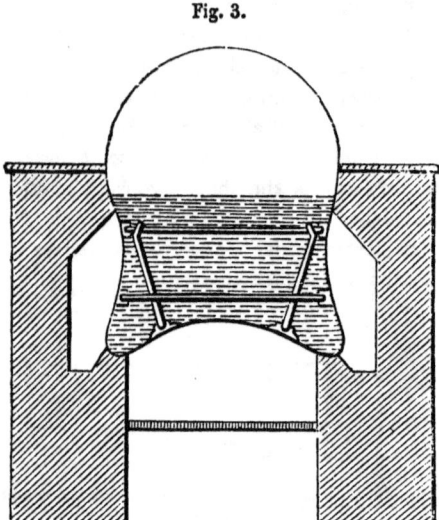

single length, each being about 2 inches thick on the upper
face, and 6 inches deep in the middle, with draft spaces be-
tween of $\frac{5}{16}$ to $\frac{3}{8}$ inch wide, according to the draft and the kind
of coal used. The grate is usually set with a fall of either
3 or 6 inches towards the back, being commonly 21 inches
from the boiler bottom at the front and 24 inches at the back
of the grate.

The boiler is always set up with a *wheel-draft*, that is, the
current of flame and smoke, after passing under the boiler
bottom, is made to rise up at the back, whence, returning
along one side by a brick flue to the front, it crosses the front
end, and then repasses along the other side again to the back
where it goes into the main flue which conducts it to the
chimney.

The upper part of this boiler is a semi-cylinder, and contains $5^2 \times 20 \div 2 = 250$ cylindrical feet, $\times \cdot 7854 =$
$196\cdot35$ cubic feet.

The lower part, if made with straight sides and flat bottom, would be equal to $4{\cdot}166 \times 5 \times 20 = 416{\cdot}66$ cubic feet; but the capacity of this part is reduced by about $\frac{1}{6}$th, owing to the sides being made concave outwards, 5 or 6 inches on each side, for the purpose of obtaining greater strength for resisting internal pressure: the bottom of the boiler being arched upwards 10 inches, or 2 inches to each foot in the width of the boiler, with a similar view, as well as with a view to equalise the action or effect of the heat radiated from the mass of burning fuel on the grate.

Therefore $416{\cdot}66 - \dfrac{416{\cdot}66}{6} = 347{\cdot}22$ cubic feet.

To this add the upper part $= 196{\cdot}35$ ditto.

$$543{\cdot}57$$

This divided by $27 = 20{\cdot}13$ cubic yards, which is the real capacity of the boiler.

Something less than half of this space is allowed for water, and the remainder for steam. Some deductions require to be made for stay-bars, straps, and cutters, which are always used for strengthening this kind of boiler, leaving about 12 cubic feet of water room for each horse power.

A rule for finding the depth of water in this boiler when the steam and water chambers are of equal capacity is as follows :—

Take half the difference of capacity between the lower and upper part, and divide it by the area of the water surface, then deduct the quotient from the depth of the lower part of the boiler, and the remainder is the depth of water, taken perpendicularly over the seating plate, at the bottom of the boiler, when the capacity for water is equal to that for steam, and which it ought never to exceed.

c

In this case—

Capacity of the lower part	.	.	347·22 cubic feet.
Ditto of the upper part	.	.	196·35 ditto.

$$2)150·87$$

Area of water surface $(5 \times 20 =)$. $100)75·43 =$ half diff.

Quotient	.	.	.	·7543
Subtract from depth of lower part	.			4·1666

Remainder in feet .	.	.		3·4123
				12

And inches .	.	.	.	4·9476

Hence 3 feet 5 inches nearly is the depth of the water.

The brickwork of the side flues is gathered in 3 inches below the surface, hence the depth of the side surface is about 3·25 feet, or measuring by the curved surface about 3·5 feet, and the total area of both sides $= 3·5 \times 40 = 140$ square feet. The total area of the two ends of the boiler below the tops of the flues is about 28 square feet less by the area of surface covered by the brick arch over the furnace mouth (about 3 square feet), and by the brickwork at the back, which divide the "uptake" from the side flue (about 2 square feet), leaving about 23 square feet, which, added to the side surface, gives 163 square feet for the total area of vertical surface; but, as we have already seen, only one-half of this can be considered as effective heating surface—it is only equal to $81\frac{1}{2}$ square feet, or little more than 9 square yards.

The area of the boiler bottom, measured by the curved surface, amounts to 94 square feet, or about $10\frac{1}{2}$ square yards, which is all effective; hence the total effective heating surface of the boiler is about $19\frac{1}{2}$, or, say nearly *twenty, square yards*.

SECTION 9.—DISADVANTAGES AND DISUSE OF THE WAGON BOILER.

The kind of boiler just described has been in such general use for so many years, and in consequence become so well known, that this collection of undisputed data in reference to it will be useful to refer to, as a standard to guide us in the estimation of the probable effects to be produced by the introduction of new forms. Up to the year 1830, when the invention of the railway locomotive created such a revolution in steam engineering, the wagon boiler was all but universal in the cotton manufactories of Lancashire, as well as being very general throughout the whole of the manufacturing districts of England and Scotland. In the year 1840 I had occasion to ascertain that about $\frac{4}{5}$ths of all the boilers at work in the immediate neighbourhood of Manchester were of the wagon shape. The additional increase of new boilers introduced into cotton factories during that interval, in order to meet the tendency to work at higher pressure, being generally either of the flued cylindrical kind already described, or of the kind called the Butterly boiler, or fish-mouthed boiler, a modification of the Cornish boiler, excellently adapted to the inferior coal and necessarily large fire grates of Lancashire. In the paper-making, bleaching, and printing works, the introduction of the plain cylindrical or barrel boiler of great length more generally prevailed; while during the same period the old-fashioned " kettle boiler," or haystack boiler of Smeaton, then so commonly used in the collieries throughout Lancashire and Yorkshire, as it is to the present day in Staffordshire, was slowly being replaced by the egg-ended cylinder boiler, which had been many years previously generally used in the collieries of Northumberland and Durham.

Since the year 1840, however, the old wagon-shaped boiler is very fast disappearing from the manufacturing districts, no new ones being now made; and as the average age of a wagon boiler when worn out may, I believe, be stated to be about twenty years, it follows that at the present time there cannot be so many left in a good enough state of repair to act as criterions of comparison. Consequently, as the science of boiler engineering is in this transition state, it may be worth while going more particularly into some details of our subject in this place, with a view to show the importance of obtaining a good proportion between the various dimensions of the boiler and the horse power of the engine it is intended to be applied to, by pointing out the particular evil effects that have generally arisen when some of the above proportions have been materially departed from.

The dimensions just given of a 20-horse wagon boiler have been frequently departed from in respect of depth, say from 6 to 8 feet, without making any great difference in the evaporating power, so long as the bottom surface and area of fire grate remained the same. Its evaporative *economy*, however, was slightly increased by increasing the depth of the side surface and narrowing the flues.

This boiler has also sometimes been made with flat sides instead of being concave outwards; but the alteration always appeared to me to be detrimental, owing to the hot air in the side flues not acting so effectually in a lateral direction as when it is allowed to impinge *upwards*, although the surface against which it acts may be ever so little inclined towards the fire. Besides, the flat sides increased the capacity of the water chamber, thus causing a waste of fuel in the first getting up of the steam, on account of the greater quantity of water required to be heated. The celebrated Mr. Murray, of Leeds, was accustomed to make his wagon boilers flat sided, and of less depth than Boulton and Watt, and gave as a reason that he merely considered the side flues as an air casing to the boiler to prevent the dissipation of heat by conduction; thus placing very little value on hot air as a steam generating agent. Yet Murray, it is well known, was eminently successful in his practice as a steam engineer, and one of the most formidable rivals of Watt; but it must be remarked that, like one of his disciples, Mr. Hick, he took the precaution of giving a so much larger proportion of bottom surface.

Section 10.—On Small Furnaces and Excessive Firing.

There are many circumstances which occur to alter both the power and economy of a boiler in the setting of it up or in the construction of the brickwork. For instance, the seating walls on which the boiler rests are sometimes made 9 inches thick instead of 4 or 5; thus reducing by so much the area of the bottom surface. At the same time, the fire grate is perhaps reduced 6 inches on each side; thereby, instead of *one* square foot per horse power, making it only about *three-quarters* of a square foot per horse; and, although this is supposed to be the proportion of fire grate allowed by Boulton and Watt, and quite sufficient, as it is, for Newcastle coal, it is notoriously insufficient for the common description of coal used in the factories of Yorkshire and Lancashire when the engines are fully loaded; unless, indeed, thin fire bars and a fire-feeding machine, or an extraordinary draft, are used.

It is certainly true that a boiler so constructed and set up with Boulton and Watt's proportions, will make sufficient steam to work a good 20-horse engine with moderate economy of fuel; but it is always found that should the engine happen to be temporarily out of order, or overloaded for a short time, so as to require one-half as much more steam than usual, such a boiler then requires what is technically known as "hard firing," the evils of which are manifold,—too numerous, in fact, to describe in any one place in this book; one of the principal objects of this work being to abate those evils by pointing out the means of removing the cause. One of the first results of two or three weeks' "hard firing" is that the boiler becomes "burnt out" in that part of the bottom which is immediately over the back part of the fire grate. Indeed, this result is always produced sooner or later when a boiler is too little for its work, and is no more than ought to be expected when we endeavour to get steam with a small fire urged to act with great intensity, rather than with a large one at a lower temperature.

In order clearly to elucidate the causes concerned in mismanaging the setting up of boilers, as well as their mismangement afterwards, we may remark that when a boiler bottom is "burnt out," as it is called, by hard firing, it is seldom or never, perhaps, that a hole is burnt through the boiler bottom at one spell; or, if so, an explosion would be the almost certain consequence. The meaning of the phrase "burnt out" is that the "nature of the iron is burnt out;" or, in other words, we find certain portions of the boiler bottom, against which the flame impinges with undue intensity, to be converted from good tough malleable iron into a state that is short, brittle, and weak. The thing generally commences and goes on in this way :—when, from whatever cause, the engine keeper finds his engine short of steam, the fireman's well-understood business is, to push the fires a little harder, which can only be done by *more frequent* firing; this, however, soon has its extreme limit in the greatest possible evaporating power of the boiler, or that point at which the production of steam would be just as much retarded by any increase in the number of times the fire-door is opened and cold air admitted, as it would be promoted by the more frequent supply of fuel. The only resource then is, larger charges of fuel at a time, constituting what is called *heavy* firing, and its attendant evil of frequent stoking. When a fire is thus urged to its greatest intensity for a few days, serious injury soon begins to manifest itself by what is called "*coming down of the boiler bottom;*" that is, the arched bottom gradually assumes a flat or depressed form about a foot in front of the bridge, and extending from side to side across the furnace, thus forming a kind of ridge projecting downwards to the fire; which, at the same time drags down into an inclined position the adjoining portions of the boiler bottom on each side of it. Now it is in the narrow strip of iron, usually not more than an inch or two in breadth, forming the bottom of this ridge or protuberance, where we find the iron permanently injured in the way described. Frequently the injury will not extend beyond a single plate, especially if the plate be a little thinner than the adjoining ones. In that case a segmental or hemispherical protuberance or dish is formed, which, as the men term it, comes rapidly " to a head," from the circumstance of its tendency to collect any deposit that may be in the boiler. Indeed, the fact of these dish-like receptacles being commonly found with dirt or other matters in them, appears to have led most engineers quite astray as to the first cause of their production, ascribing their first formation to the accidental presence of such matters instead of looking on them as the effects produced, or at most secondary causes only.

The true theory of the matter has always appeared to me to be very plainly pointed out by the fact, which may be safely assumed, that whenever a plate is thus overheated, the generation of steam has been so rapid over that particular place as in a great measure to drive the water away from the surface of the plate; or at least such a degree of repulsion of the particles of the water is effected as to prevent that immediate and intimate contact which is necessary to keep the iron as cool as the water over it. Now the natural sequence to this state of things is easily inferred; not that the iron will get *red* hot, perhaps, which is in fact not required to produce the phenomenon in question, but examinations of the iron in those cases generally show that a *blue* heat has been produced, a temperature something exceeding what is called the maximum evaporating point, and considerably exceeding the temperature of the iron necessary to effect the repulsion of the water we are speaking of, which has been stated by M. Boutigny, of Evreux, in France, who has taken great pains to investigate the subject, to be as low as at 350° of Fahrenheit; while the water in the boiler may be only 250°, or less. This repulsion may continue for a few seconds only, while the softened plate is exposed to the pressure of the steam forcing it down in the way described: at the same time, from the

agitation and boiling of the water, the heated part is constantly exposed to the washing of the water over it; thereby
alternately heating and cooling the part which we eventually
find " burnt out."

SECTION 11.—GENERAL MISMANAGEMENT OF BOILERS AND TRADE OF BOILER MAKING.

It most frequently happens, when a boiler bottom is
burnt out in the manner above stated, that the proprietor,
instead of causing an investigation into the circumstances of
the whole case, including his own orders perhaps to overload
the engine or to use an inferior kind of coal, immediately
sends for the boiler maker, or rather boiler mender, when,
after some talk about bad iron, bad or dirty water, and being
short of water, the injured plates are cut out and replaced
with new ones. The bricklayer replaces his bricks and
mortars precisely as before, when, as the same causes continue,
of course a similar rupture takes place before another month
is over ; again the same tinkering process is resorted to, and
after a few rounds of this kind, when the expense of repairs
has amounted to nearly as much as might have purchased a
new boiler, to say nothing of the trouble and confusion as
well as the interruption of work necessarily attending such
jobs in a regular factory, the wearied proprietor at last
sends for another boiler maker, the old boiler is then con-
demned and a new one ordered in its place. All this takes
place simply for want of removing a few unnecessary bricks
from the sides of the fire-place, and taking care not to put too
many coals on the fire at a time.

Frequently a 30 and sometimes a 40-horse boiler is ordered
to replace one of 20-horse power. But when the new boiler
is to be set up, it commonly happens that the same want of
knowledge is evinced by both boiler maker and bricklayer as
in the case of the first. Occasionally I have known the very
same fire grate used to the new boiler that caused all the

trouble at the old one. Nay, I have known instances where a *smaller* grate has been recommended by first-rate engineers to a larger boiler, on the ground that the increased quantity of heating surface would render a less quantity of fuel necessary, and therefore a smaller grate ;—which appears feasible, but is only partly true on the supposition that the grate had been previously so large as to make it difficult to keep it covered with any smaller quantity of fuel, which is contrary to the case supposed.

There is one thing which cannot be too strongly impressed upon the minds of engineers who are called on to give opinions in difficult cases of this kind. It is that the evaporative power of a boiler ought not to be confounded with its evaporative economy ; the former corresponding to the quantity of steam it is capable of producing without injury from over-firing, irrespective of its economy, and is entirely dependent on the area of the fire grate; which, other things remaining the same, limits the utmost evaporative power of the boiler.

When a large boiler is set up in place of a small one, any saving produced by the greater extent of heating surface is, in factories, often counterbalanced by the extra consumption required in getting up the steam every morning, on account of the larger quantity of water it contains. Generally speaking, if there is a sufficient supply of steam while the boiler is at work, all parties are satisfied; and although the consumption of fuel may be larger than it ought to be, it is seldom found out immediately, or if so, it is easily laid to the account of the newness of the brickwork, and before correct opinions can be formed of it, many other changes are liable to occur to take the blame from the unnecessarily large boiler ; of instances that have occurred within our knowledge we may enumerate a few; namely, the fireman may have been changed—the coal has been obtained from a different quarter—some alteration has again been made in the engine, or in its load—the oil for the machinery driven by the engine is different—the

weather may be colder and require more steam to heat the
factory, or it may be wetter, and the atmospheric pressure
low, consequently the draught of the chimney will be worse
than usual, nay, the weather may be too warm where there
is a scarcity of condensing water, and consequently the vacuum
of the engine will be impaired ; and in short, where there are
so many circumstances liable to daily variation, it rarely
happens that the exact state of the case is ascertained even by
those immediately interested.

It may, and indeed is, very frequently said that it is always
advisable to have plenty of boiler room at the commencement,
even if the consumption of fuel should be increased 5 or 10
per cent. ; which is certainly not much in a small concern, say
20 to 40-horse power, when set against the convenience of
having it in your power at any time of adding 8 or 10-horse
power to your engine, by putting in a larger cylinder.   But,
if we consider the setting out of new establishments upon the
immense scale of many recently set to work in the cotton
spinning district, where the expense of fuel frequently amounts
to two or three thousand pounds per annum, a saving of only
5 per cent. is not a trifle to be overlooked; to say nothing of
the absurdity of laying out large capitals upon mere guess-
work principles, or at least in such a manner as to make
extensive alterations in the boilers and engines a matter or
almost certain expectation before the concern is fairly at work.
So far at least as the size and number of the boilers are concerned, it is a notorious fact that a great majority of the
newly-erected cotton works in Lancashire have had to undergo
such alterations.

When the capitalist or manufacturer is about to erect new
works whose moving power is steam, unless he be a man of
sufficiently comprehensive mind to have a general practical
knowledge of the various trades of builder and bricklayer,
engine and boiler maker, millwright and machine maker, as
well as that of whatever business his factory is designed for,
let him be his own architect and engineer, and it will be in

vain for him to expect anything like a scientific execution of his plans, although his general design may be ever so judiciously arranged.

The two trades of engine and boiler maker are usually combined in the same party, but in Lancashire it is far from being general, except in those establishments that are very extensive; boiler making being now considered nearly as a separate trade, which circumstance, perhaps, as much as any other, was the cause of the present investigation of the subject.

There is one custom which has a great tendency to retard improvement in this branch of business, and that is the prevalent one of contracting for the making of boilers by weight instead of so much per horse power, as is the practice of the engine maker. This custom is general, and is a direct inducement for unprincipled makers to compete with each other until there is not enough profit left to repay even so little skill as is required to make the most ordinary rough draft on paper of the article wanted.

When a manufacturer wishes to give an order for a new boiler, he commonly first ascertains the price of iron per cwt., and also the rate of wages per cwt. for making, and he has then only to find out some master boiler maker who will admit of being squeezed down to the minimum of profit, and he may then expect the natural consequences of competition; namely, a much larger boiler than is necessary, besides being a great deal heavier than it ought to be even for that large size. A heavy boiler has also other faults that are more serious, arising from the difficulty of riveting by mere manual labour very thick plates of iron, so as to be perfectly watertight, without the use of various cements to prevent leakage, such as red lead and iron borings. There is also a very reprehensible practice of making use of a wash for the purpose of *rusting* the iron in the imperfectly filled rivet-holes and unclosed joints of the plates. By using a solution of sal-ammoniac for this purpose, a coating of oxide or rust is given to the inside of a boiler, in a few days, sufficient to prevent leakages until the boiler has been a short time at work, when a little dirt or incrustation comes to assist the rust in forming a more permanent covering. This, when once formed, becomes a regular lining or shield of non-conducting substance between the heating surface of the boiler and the water which is to be heated. Thus it is that the unsuspecting manufacturer comes to be furnished with what he considers at the time to be a *cheap* boiler.

SECTION 12.—ON ENLARGING THE FURNACE AND LENGTHENING THE BOILER.

When a wagon boiler is found to have insufficient power, it frequently admits of a very effectual remedy by lengthening the fire bars : and when the seating walls are unnecessarily thick, by widening the furnace throughout. A great portion of the brickwork on each side may be removed by supporting the boiler on six or eight cast-iron blocks or short columns, and with merely a brick in breadth wall to divide the flame-bed and furnace from the side flue.

When this plan is done carefully, it is always followed by a

Fig. 4.

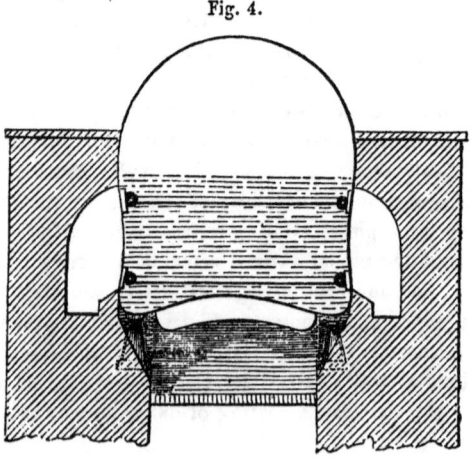

great increase in the evaporating power of the boiler, without requiring any great addition to the area of the fire grate.

In effecting this alteration care must be taken that the enlargement of the fire grate does not injure the draft, otherwise a contrary effect to that expected has sometimes been produced, particularly where the chimney is small and no surplus draft at command. In such a case it is necessary to diminish by a small amount some of the spaces between the grate bars, in order that the total area of draft space may not be increased in so great a proportion as the area of the grate itself. The easiest way of effecting this is to chip or plane off one side of the heads of each of about half a dozen bars at the sides of the grate, leaving the largest draft spaces towards the centre of the fire; and if there are two lengths of bars, confine the operation to the first length, by which means the combustion will be more rapid where otherwise the coals are apt to accumulate.

It has been already remarked that the evaporating power of a boiler is always found to be, other things remaining the same, in proportion to the area of the fire grate; and to this may be added, the evaporative economy of a boiler is always much increased by any increase of heating surface immediately over or very near the fire, the area of the fire grate itself remaining the same, or in some cases even diminished.

There is another fact connected with this part of the subject which must have struck any one who has been at any considerable pains in making observations, which is, that the heating surface very near to, or over, the fire grate is so much more effective than those portions of the boiler which are beyond the occasional reach of the flame, as scarcely to admit of any comparison. Certain it is, that 2 or 3 square feet of additional surface over the hottest part of the fire grate makes a considerable improvement in the power as well as in the economy of a boiler, while as many square yards of surface added to the contrary end has scarcely any perceptible effect at all, provided that the proportions of the boiler in other re-

spects remain the same.   It has been found that when boilers
of 20 feet long have been lengthened by 10 feet, the addition
has made scarcely any difference in the evaporating power,
while the fire grate remained the same, and the saving of fuel,
if any, has only been very trifling.   It ought, however, to be
stated, that those instances were confined to the generally open
burning coal of Lancashire, and to the Lancashire practice of
thin firing and *no stoking;* and it may be doubted whether
the same results would be obtained where a more bituminous
coal, together with a barbarous practice of stoking, are used.
In one particular instance I recently met with in that county,
a large boiler, constructed on the Cornish principle, of 30 feet
in length, was made 15 feet longer with very little or no improvement in either power or economy.   Again, I have seen a
boiler cut down from 104 feet in length to *fifty,* without any
sensible diminution in its evaporating power.   In this case,
both before and after the alteration, the temperature of the
air and smoke escaping into the flue of the chimney was so
considerable as to cause water to boil in an open vessel very
readily.

Those who are not aware of the great extension of steam
power in Lancashire during the last twenty years, may be sur-
prised at the existence of steam boilers of such a large size as
the one just mentioned; I may therefore state that the instance above mentioned occurred at the extensive paper works
of Messrs. Crompton, of Ringley Bridge, near Manchester;
and that several boilers of 60 and 70 feet long may be met
with in the same neighbourhood.   There are also a few instances of very large boilers in factories; one boiler in particular, worked some years in a cotton mill in Manchester,
which may be mentioned as probably the largest boiler in the
world: it was a cylinder of 8 feet in diameter and 90 feet long,
with a flue tube running through its whole length; and if it
could have been properly fired, would have been about 200
horse-power.

CHAPTER III.

On Proportioning Surfaces, Capacities, and Fire Grates, with Rules and Calculations.

Section 13.—Propriety of adopting a Square Yard of Surface for each Boiler Horse Power.

From what has been already stated there cannot be a doubt that an effective heating surface of *nine* square feet in area, together with *one* square foot of surface of fire grate, is in any case quite sufficient for each horse power.

It appears also from the two practical examples given (sec. 5–7), that whilst one square foot of fire grate per horse power was barely sufficient under difficult circumstances, and with the inferior coal of Lancashire, it was *more* than sufficient under similar circumstances in London by just so much as the Newcastle coal used in the latter case was superior to the former. Yet in both cases the proportion of one square yard of heating surface per horse power appears equally eligible ; giving as it does a proportionately greater extent of heating surface per unit of fire grate, as the heat-producing power of the coal used in one case exceeds that in the other.

Where it is expedient to fix on a convenient round number for a *nominal* measure of effects expected to be produced by similar causes acting over certain areas of surface, as the fire and flame do against the bottom and sides of a boiler, it is not easy to conceive more reasons all pointing one way than in this case point to a square yard as the natural unit or measure of a steam-boiler horse power.

Although this unit measure accidentally occurred to us at the commencement of our steam-boiler experiments in the way previously described, we happen to know that it has been generally approved by others ; and as no one else has proposed any other, that we have heard of, we have the less hesitation in finally adopting it in this work.

The convenience to practical men of adopting round numbers as unit measures, when they come pretty near the actual values, in all results that only admit of being estimated, is so great, that some, perhaps, may be induced to suspect that we have strained the facts a little on purpose to make them fit the rule. If so, it is only necessary to mention that the rule has been published for more than a dozen years past, whilst the remarkable coincidence of the two examples referred to in corroboration of its being at least near the truth, is only of very recent occurrence; one of the boilers having been made within the last few months. Besides, it may be sufficient to state that general rules and proportions, closely approaching the above, have been notoriously in use a great number of years, and the facts corresponding with them are easily proved.

Although any simple methodical system of stating the combined results of calculation and observation is highly to be valued, as being conducive to the ready apprehension and easy application of most useful knowledge, we ought not to be such sticklers for mere methodism as to risk the smallest deviation from truth, when the latter admits of correct ascertainment. In physical problems, however, especially of the nature of those now treated on, there is an absurd refinement of figures and fractions too frequently to be found in the reports of some of our railway and other engineers. This is especially the case in recording series of observations, for the purpose of obtaining averages, or striking a mean in the ordinary way, which may be easily tested by any one, when the useless exactness of minute fractions, while there are errors in the integers, becomes immediately evident. This ultra exactness increased greatly with the railway mania, and can be useful in nothing, saving perhaps " cooking accounts," for it practically produces more error than the actual errors of observation themselves are generally capable of doing.

For the above as well as other reasons which could be adduced, engineers generally are respectfully urged to sanction either this or some other measure, equally feasible and convenient, as the unit of steam *boiler* horse power. Like the steam *engine* horse power, it is not at all necessary that it should be an absolute exact average of the power of a living horse, but as near an approach thereto as to enable any one not conversant with the details of the subject to have a ready apprehension of its meaning as an *estimate of power*, and at the same time a tolerably correct idea of the force capable of being produced by the boiler indicated by it. For this purpose it is also expedient that, whatever difference there is between the calculated *boiler horse power*, and the *real horse power*, the error should be in excess on the side of the boiler, as is the case with Watt's *steam-engine horse power*, which is, as every one knows, much more than the strongest horse can do for any considerable length of time.

This standard of Watt's for the steam-engine horse power has been vainly endeavoured to be altered; but the comparative incongruity of all other measures that have been proposed have been at once seen and repudiated.

SECTION 14.—PROPORTIONING THE FIRE GRATE TO THE HEATING SURFACE AND THE QUALITY OF THE COAL.

As has already been observed (sec. 12), when a boiler does not produce a sufficiency of steam, we have in many cases only to cut away the side walls of the furnace, and thus, by increasing the width of the fire grate, considerably augment the power of the boiler. This simple and inexpensive plan of enlarging the furnace and fire grate has been carried out in very many instances to the extent of increasing the evaporating power of a boiler to nearly double that which the quantity of heating surface would otherwise warrant, and without any very material increase in the ratio of the fuel consumed to the water evaporated.

There are also abundant opportunities of proving that, with the ordinary engine coal of Lancashire, it is difficult to keep steam with less than a square foot, and with Newcastle coal three-quarters of a square foot, of fire grate per horse power, whatever amount of heating surface there may be; which, taken at one square yard per horse power, gives us in the former case the proportion of *nine*, and in the latter that of *twelve* square feet of effective heating surface to each foot of fire grate.

It is not meant to be denied that a greater proportion of heating surface to fire grate than either 9 or 12 to 1, which we have assumed, would be more economical in theory, but it would be so in theory only, and not in practice. Indeed, if we follow theory only, it may be doubted whether a ratio o. 100 to 1 may not be nearer the correct proportion, for it is evident that the hot air or smoke ought not to pass into the chimney until it is deprived of its power of generating any more steam; or, in the language of the chemist, "until it has imparted the whole of its superabundant caloric to the water," always excepting so much as is required to cause a good draft in the chimney.

In accordance with correct scientific theory, the hot air ought to be kept in contact with the boilers until reduced to 212°, or to such a temperature as corresponds with steam of the required pressure, which for a low-pressure engine would seldom be much above 220°. This may appear to some to be a good point of theoretical perfection to be aimed at, but it is a very fallacious one; and the greatest possible absurdities have been frequently committed by attempting it in practice. There is nothing to be obtained by any extension of the heating surface beyond certain limits, which fall very far indeed short of the point mentioned. In fact, I have met with very few instances of *successful* practice where the temperature of the gases leaving the boiler has been less than 400° too 500°, whilst I have found many cases of very great economy, both with the Lancashire and the Newcastle coal, where the temperature of the smoke (and black smoke too) passing into the chimney has been equal to melt-ing lead, or above 600°.

Now, when I say that nothing is to be gained by an ex-tension of the boiler surface beyond the limits indicated, I mean that the saving would be less than the simple interest of 5 per cent. per annum upon the cost of the extra boiler-plate, and other contingent expenses required for such addi-tional extension of heating surface. The small saving of fuel is admitted, but in practice it costs more than it is worth; and herein engineers may see an example of many cases in which theory not founded upon experiment and observation is at variance with practice; the scientific theorist will not admit the truth of anything which there is not a good reason for, while the practical man, quite as dogmatically, but with better reasons for being so, refuses to adopt any plan which has not been found to answer, whether there is a reason for it or not.

It is evident that there will be a different ratio of heating surface to fire grate, suitable to the different qualities of fuel, but we are now speaking of the average quality of engine coal used in the manufacturing districts. With that, and the fire managed in the ordinary way, the result of my experi-ence is, that 9 square feet of effective heating surface for each square foot of fire grate is sufficient to enable us to make such an arrangement of the forms of all the simpler kinds of boilers as in all cases to reach the full limits of real practical economy.

The best engine coal about Manchester is generally just as it comes from the pit, containing all the small and inferior coal, dirt, &c., which is usually left in the mine when house-fire coal only is wanted. The worst coal is commonly called "Burgey," and is of various qualities; it generally consists of the tops and bottoms of the seams, together with other refuse coal that is left after the house-fire coal is mined. It does not require a larger fire grate than the best, but it is found

to require a much better draft.   It is not so good as "slack,"
which is riddled from house-fire coal, especially when the latter
is put on by the firing machine.

The best Newcastle coal supplied to the London market for
engine purposes appears to me to be the different Hartley's ;
which I find requires about three-quarters of a square foot of
grate for each horse power at the same time ; a square yard of
effective heating surface per horse (or in the proportion of 12
to 1) seems also capable of absorbing the heat produced sufficiently to reduce the temperature of the gases, leaving the
boiler to the same degree that was found most economical in
Lancashire, namely, below the melting point of lead.   It is
lighter and burns well with less draft than the best Lancashire
coal.

For Welsh coal and some others the fire grate will bear a
smaller ratio to the heating surface ; and for coke the least
of all.

Whatever may be the amount of heating surface in a steam
boiler, the quantity of steam to be obtained from it will of
course depend entirely on the area of the fire grate being proportioned to the quantity and quality of the coal which has to
be consumed upon it.   Consequently, if we find a boiler mani-
festly too small for producing the quantity of steam that may
be required, the object is at once accomplished by enlarging
the fire grate in the way already pointed out, when any quantity of steam may be obtained that the draft of the chimney
will admit of, although with a less degree of economy in fuel ;
therefore such alterations are only to be recommended in urgent
and exceptional cases.

In practising this method of increasing the evaporating
power of boilers, I always found that the maximum effect was
produced when the area of the fire grate was increased in a
somewhat greater ratio than the effective heating surface was
diminished ; that is, if a certain effect is to be produced, say,
for example, an evaporating power equal to the supply of a
20-horse engine, which we have seen requires 20 square yards

of surface, but if from mal-construction of the boiler or other circumstances we have only 18 yards, then the furnace for burning Lancashire coal will require to have a fire grate of *rather more* than 22 square feet in area; or should the heating surface be only 16 yards, then the fire grate will be required to be *more* than 24, in fact about 25 square feet.

It is not that any such great nicety is always required as is set forth in the above proportions. For in effecting alterations for increasing the power of the boiler that is short of steam, the best way is to get all the increase we can, and if too much, trusting to the damper as the most profitable means of diminishing it. But after an alteration is effected, it is sometimes of use to know at what rate we ought to work the boiler at, in order to produce the greatest economy it is capable of under the circumstances. This is more especially of use when a number of boilers of different powers are working in conjunction, in order to avoid working too hard, and thereby, as it commonly happens to the most willing horse in a team, over-firing and injuring perhaps the best boiler of the set.

For this purpose I have been in the habit of using the following empirical rule, which is derived from the observation already made, that in the case of a 16-horse boiler having a 25-horse fire grate, it is always safer as well as more economical to work it rather below than above the *arithmetical* mean between those two numbers $(20\frac{1}{2})$; therefore the *geometrical* mean (20) was adopted. An additional reason for adopting the geometrical rather than the arithmetical mean, may also find favour with some who, like myself, may have no abstract love of squares and square roots,—it is the fact, that the operation on the slide rule for finding any of the three numbers in question is in the former case the simplest of the two; as we may now proceed to show.

SECTION 15.—RULES FOR FINDING THE HORSE POWER, THE AREA OF FIRE GRATE, AND THE AREA OF EFFECTIVE HEATING SURFACE, USEFUL IN ALTERING OR RESETTING OLD BOILERS.

Let s = the effective heating surface in square yards.

    F = the area of the fire grate in square feet.

    P = the horse power of the boiler.

Then, using the inverted slide-rule or common rule with the slide reversed, it must be set as follows:—

A	Horse Power = P	Fire grate in sq. feet = F
Ɔ	Horse Power = P	Heating surface in sq. yards = s

Suppose, for example, that we have 16 square yards of effective heating surface in a boiler, with a fire grate of 25 square feet, and we wish to know the most suitable rate to work the boiler at. Then place s = 16 on Ɔ, against F = 25 on A, and looking to the left hand, the first two divisions of the same value that coincide with each other (which are those of 20) represent the horse power, as in the example below.

A	P = 20	F = 25 sq. feet.
Ɔ	P = 20	s = 16 sq. yards.

Again, suppose that a boiler is intended to drive a 30-horse engine, but it has only 25 square yards of surface, and the area of the fire grate is required. Then place 30 = the horse power on Ɔ, against the same number on A, and looking to the right hand, you will find, against 25 yards of surface on Ɔ, the answer to be nearly 36 square feet upon A for the area of the fire grate; and in like manner against any number expressing the square yards of effective heating surface on Ɔ, you will find the number of square feet of fire grate most suitable for the given power upon A. In short, the inverted line c represents a table of 30-horse boilers with various quantities of surface, and the line A represents a table of areas of fire grates corresponding to those surfaces.

A	30	Fire grates	31	34	36	39
ɔ	30	Surfaces	29	26½	25	23

The rule, when thus set, exhibits at a glance the various ways in which a 30-horse boiler may be set up.

It will be seen that any one of these three principal data respecting boilers may thus be found when the other two are given, by any workman in the possession of a slide rule; and it is for his use only that these examples are given.

As rules and examples of the arithmetical operations for the foregoing questions may be useful to a large class of workmen who may wish to exercise themselves at figures, they are also added,

RULE 1. To find the maximum horse power at which to work a steam-engine boiler that is not well proportioned.— Multiply the area of the fire grate in square feet by the area of the effective heating surface in square yards, and the square root of the product is the horse power required; provided always there is not less than 1 square foot of grate for each yard of surface.

Example 1. Suppose a boiler to have 16 square yards of effective heating surface, and a fire grate of 5 feet square, or 25 feet in area: required the maximum horse power at which to work the boiler with *Lancashire* coal, in order to have as little waste as possible.

Area of fire grate . . . 25 sq. feet.
Ditto of surface . . . 16 sq. yards.

$$\begin{array}{r} 150 \\ 25 \end{array}$$

Answer.

Product . . 400(20-horse power.
400

$$\overline{000}$$

Now if, instead of Lancashire coal, as in the above example, we suppose the same boiler to be worked with either the best *Welch* or with *American* coal, which we may suppose, for comparison sake, to be of *double* the strength of the *Lancashire* coal; or, in other words, that instead of requiring a square foot of grate per horse power we find half a square foot sufficient, —then the same question may be stated in this way :

Example 2. A boiler having 16 yards of surface, has a 50-*horse fire grate*, required the power it may be worked at without waste.

Area of heating surface .   .   16 square yards, as before.
Horse power of fire grate   .   50
                       —— *Answer.*
                  800(28·28 sq. root, or horse
                   4   power required.

         48)400
            384

       562)1600
          1124

    5648)47600
         45184

In the same manner, similar operations may be performed for other fuels, using the number of horses' power that the fire grate would be equal to according to the value of the fuel used, instead of the area of the grate itself.

Rule 2. To find the proper area of fire grate most suitable for working a steam boiler, with a deficient heating surface, in order to drive an engine of any given number of horses' power.—Divide the square of the given horse power by the square yards of effective heating surface, and the quotient is the area of the fire grate in square feet for *Lancashire* coal.

When any other coal is used of a different quality, then divide the last quotient again by a number corresponding with the quality or value of the coal used, when the Lancashire coal is called 1, which will give the proper area of the grate in square feet.

Example 3. Suppose a steam boiler has 16 square yards of effective heating surface, what must be the area of the fire grate to enable it to drive a 20-horse engine with the greatest economy in Newcastle coal?

$$\text{Horse power} \quad . \quad \frac{20}{20}$$

$$\text{Surface, } 16 \begin{cases} 4)\overline{400} \\ \overline{} \\ 4)\overline{100} \end{cases} \text{ sq. of the horse power.}$$

$$\text{Square feet} \quad \overline{25} \quad \text{area of fire grate for Lancashire coal.}$$

Again, if we consider Newcastle coal to be one-third better than the average of Lancashire taken as unity, its value will be expressed by $1\frac{1}{3}$ or $1\cdot33$; therefore,

$$
\begin{array}{r}
1\cdot33)25\cdot00(18\cdot7 \\
13\ 3 \\
\hline
11\ 70 \\
10\ 64 \\
\hline
1\ 060 \\
931 \\
\hline
\end{array}
\quad \text{area of fire grate for Newcastle coal.}
$$

Section 16.—On the Capacity of the Steam Chamber.

Before going further into the general principles that govern the form and construction of the steam-engine boiler, we have an element to take into consideration of not much less importance than the heating surface already disposed of, and that is, its capacity for steam and water, but more particularly for steam.

The proper capacity of the steam chamber or steam chest, has been much less considered by engineers than the area of

D

heating surface. Indeed both theoretical and practical men have generally agreed in this matter, as they occasionally do when they are both wrong, to think it of little consequence. We shall, however, soon see what blind guides a *little* reasoning upon a limited observation of facts is capable of making.

If, instead of the ordinary steam chamber of a boiler, we have a detached steam chest or close box or vessel, containing a permanently elastic gas, or even steam itself separated from water, allowing its temperature to remain the same, and supposing a communication to be made, by means of a cock or valve, with a second vessel of the same capacity as the first, and quite empty, then, as soon as the second vessel becomes filled with the expanded steam, it is evident that by occupying double the space the elastic force of the steam will be reduced exactly one-half; the resulting pressure being always inversely as the expansion. Now, if we further suppose the first vessel to be *constantly* receiving a uniform supply of the steam, we can obtain a rule for making the two vessels in just such proportion to each other as will keep any variation in the elastic force of the steam, in the first vessel, during any given short periods of time required for shutting and opening the valve of communication, within any given limits.

There is no need, however, to follow Tredgold's example by giving the calculation, because the two vessels are very far from being a parallel case to that of a steam-engine boiler and cylinder. This is owing to the fact, which appears to have been quite unknown and unnoticed, both by Tredgold and all other writers, that *there is nothing like a uniformity of supply in the boiler;* on the contrary, the supply of steam from the water in the boiler is *more violently intermittent,* if the expression may be allowed, than any mere opening and shutting of valves can account for.

Although the effect of the action of the fire against the boiler may be supposed to be uniform, yet by far the greatest portion of the steam is very rapidly generated during the first half of each stroke of the piston, and at which time only ebullition, properly so called, takes place, but which usually ceases immediately after the piston passes the point of its greatest velocity,—and the great difference in the quantities of steam supplied by water *in a boiling state* and when *not boiling*, although at the same temperature, it is unnecessary to insist on.

The steam in a boiler continues of course to accumulate, although slowly, during the time that the induction valve is quite closed; but as soon as the momentum of the fly-wheel carries the crank over the centre and opens the valve, the surface of the water in the boiler is, with the commencement of each stroke, simultaneously relieved from a portion of the pressure of the steam, and the water immediately commences boiling, let its temperature be what it may.

A very demonstrative way of illustrating this intermittent action of the water of a steam-engine boiler, is to adapt a test tube with a small portion of the bottom cut off, to the neck of a Florence flask containing boiling water, and placed over an argand lamp or gas-light; then by using a cork by way of piston, and the test tube as a cylinder, the phenomenon we are speaking of, may be produced in miniature exactly as it occurs with the steam engine itself; and with this great advantage that it can be *seen*, and therefore easier understood.

It will be perceived by those who may repeat this experiment, that the manner in which steam is produced in the boiler of a steam engine *at work*, is partly by a *distillatory* process during the times that the valves are closed, or in the act of closing, or, properly speaking, by *evaporation*, which is dependent upon the extent or superficial area of the evaporating surface or surface of water, as well as upon that of the heating surface; but of course the steam is principally produced during the short period of violent ebullition, its production in the latter case being termed by chemists *vaporisation*, which seems to mark a proper distinction; we shall, however, generally use the more common term *evaporation*, to express the compound process we have been describing.

D 2

Returning to the assumed case of the two vessels, and sup-posing that the first (which may represent the steam chamber of a boiler) is constantly receiving a uniform supply of steam, equivalent to Tredgold's idea of a constantly-uniform evapora-tion of the water, then it would evidently follow that the pres-sure of the steam must always retain the greatest degree of uniformity when the flow of the latter through the channel of communication with the second vessel (which vessel may re-present the cylinder) is interrupted for the shortest possible period of time; and consequently under similar circumstances a smaller capacity of steam chamber would be sufficient to render any variation of pressure inconsiderable, than if those interruptions were to occupy a greater portion of time. A precisely similar conclusion, founded upon the mistaken as-sumption of an equability in the generation of steam, is the groundwork of the erroneous theory of Tredgold, as may be seen by reference to his work (art. 210, &c., original ed.).

The above error once adopted led naturally enough to the conclusion arrived at by Tredgold, after a mathematical in-vestigation of the point, that *an engine working expansively requires a larger steam chamber than when acting at full pressure;* namely, that in a double-acting engine, with the steam cut off at half stroke, the capacity of the steam chamber should not be less than *eight* times the volume of the steam used for each stroke, whilst for the same engine, working at full pressure, he states that about *three* times the quantity used for each stroke may be sufficient.

The impossibility of such a conclusion being the true one, would have struck any one practically acquainted with the subject, from its absurdity, which is only equalled by a similar statement, quoted by Tredgold from Prony's "Architecture Hydraulique," to the effect that it is one of the advantages of a double-acting engine that it *requires a smaller boiler than a single-acting one;* which observation must have arisen from the same false premises and the same mode of reasoning.

If, as according to the above doctrine would be the case, all those engines whose valves are closed for the shortest portion of time between the strokes required the least capacity of steam chamber, then the old-fashioned hand-gear or tappet-valve engine, in which the action of shutting and opening the valves is effected the most suddenly, would be able to do with less steam room than the modern slide-valve engine, which is worked by an eccentric; but it is proverbially well known that the reverse of this is the case.

With respect to the proportion that Tredgold finally recommended in practice, it is fortunate perhaps that he met with a remark of Dr. Thomas Young's, to the effect that the steam chamber should contain *ten* times the volume of steam consumed at each stroke; for, after quoting Millington's opinion, that *five* or *six* times, and giving his own that *three times*, that quantity is sufficient (which is equivalent in a low-pressure engine to about 3 cubic feet of steam room per horse power), he appears to have thought it " one of those cases which seems to be incapable of being investigated, otherwise than by experience," and ends by recommending, that " to limit a low-pressure steam boiler of a double-acting engine to a change of elastic force not exceeding $\frac{1}{30}$th, we must have *ten feet of space for steam*, and ten for water, for each horse power."

Although there are many boilers working with this proportion of steam room, and even with a smaller proportion, all of them that are capable of doing their work with moderate economy, are either liable to *prime*, or belong to engines in which the steam is very much *throttled*, or *wire drawn*, the pressure being invariably a great deal higher in the boiler than in the cylinder, and in some cases nearly double. From this cause, mainly, we have the tampering with the safety valves, and consequent accidents continually occurring where high-pressure engines are in use. In low-pressure engines this method of meeting the difficulty of too small a steam chamber, by raising the pressure of the steam, is not practicable to anything like the same extent, on account of the usual method of feeding the boiler with water by means of the ordinary stand pipe, at which the water boils over when the steam is too high.

Repeated observation for years has convinced me, that for working an ordinary condensing engine in a factory, where, until recently, the steam in the boilers was seldom more than 2 lbs. per square inch, greater pressure than the average pressure on the piston throughout the stroke, no less than 12. *or* 14 *cubic feet of steam room* per horse power is required, to enable the boiler to do its work properly. On that account I do not hesitate to recommend the steam chamber to contain at least half a cubic yard for each nominal horse power.

Section 17.—Causes and Effects of Priming.

The proportion of half a cubic yard of steam room, under ordinary circumstances, as a minimum for each nominal horse power, is recommended as the result of no very limited experience, being from actual personal examination and measurement of several hundred boilers, together with the recorded indicated power and economy of the engines to which they were attached, principally with a view to the determination of this particular datum.

Instances are occasionally met with where double the above proportion of steam room is not sufficient to prevent the boiler priming; but there are in such cases always some special circumstances to account for it. For instance, where there has been any neglect in cleaning the boiler; but it fortunately happens that when a boiler primes from this cause, it not only gives both visible and audible signs of its taking place, but also leaves in some measure permanent traces of its having done so, sometimes in the cylinder, but almost always in the boiler, so that the accuracy of statements on this head admit of being satisfactorily tested.

The presence of some kinds of dirt in the water, particularly if it be of a mucilaginous nature, is liable to cause.

the engine to prime, whatever may be the amount of steam room. A small quantity of soap has a wonderful effect in this way ; and, generally speaking, with a dirty boiler and a full load on the engine there is a continual "flushing" of the water in the boiler, or an *attempt* to "prime" at every stroke of the engine. In such cases we always find that the dirt has partially left the boiler and passed into the cylinder, thereby seriously injuring the packings of the piston and valves, creating a great deal of extra friction, lowering the speed of the engine, and causing great consumption of tallow and fuel.

It is a remarkable fact, that when the inside of a boiler is examined immediately after it has primed much from this cause, marks are usually found running in a slanting direction up the sides of the steam chamber towards the steam pipe in nearly a direct line, leading from that part of the surface of the water which is over the hottest. part of the furnace, just as if a partial explosion (if we may so call it) of the water had taken place in that part of the boiler, and spent its main force in the direction stated. When there is any considerable quantity of mud or clay in the boiler, the marks indicating the direction of this explosion or "flushing" are particularly legible, in streaks or ridges of mud reaching far above the water service; and the effects of the violent agitation of the water in such cases, I have frequently observed in the forcible removal of heavy articles, which had been placed in the boilers for the purpose of experiment.

This peculiar tendency of the water in a steam-engine boiler to rise into the cylinder of which we are treating is well known to all experienced *operative* engineers, under the name of "priming," "flushing," "pumping," and other similar terms, and is known to many as being the principal, if not the only thing which makes a large steam chamber necessary ; for under certain circumstances connected with deficient capacity of steam chamber, the engine is found to prime, however clean the water may be in the boiler.

In analysing the various causes concerned in the production of this effect, it is necessary to take into consideration the effects produced by the necessarily intermittent action of the valves, referred to in the last section. The supply of steam to the cylinder being momentarily cut off at the end of each stroke, as well as for a considerable period during each stroke, in those engines that work expansively, the effect is to throw the water of the boiler into a slight undulatory motion, as may frequently be observed of that in the glass-tube water gauge. This undulatory action of water in the boiler, although caused by and dependent on, appears to be in addition to, the intermittent state of alternate ebullition and repose before described. At any rate, from one of these conditions or from both combined, it is certain that the motion of the water occasionally becomes so violent as to carry it in a state of spray or foam into the cylinder along with the steam, and when in too great a quantity to escape through the steam port in the return stroke, it infallibly breaks down the engine.

This effect must inevitably follow the priming of a sufficient quantity of water in the cylinder of a beam engine in a factory, because it is not and cannot be expected to be calculated to withstand a sudden *blow*, and such it is in reality. For if the water primes into the cylinder in the down stroke, it must remain on the top of the piston until it strikes against the cylinder cover in the up stroke, with more or less violence according to the quantity. From the incompressibility of water, the effect is the same as if a piece of iron of equal thickness to the depth of water on the piston was suddenly inserted in its place. The tremendous effect sometimes produced when a large engine breaks down from this cause, may easily be conceived; for as the vacant space left for clearance at the top of the cylinder is generally about the same depth in large as in small engines, the intruding body of water strikes the cylinder cover with a proportionally greater force. Generally the accident does not end with merely straining or breaking the crank pin, which may be the extent of the injury in small engines; but the momentum of the beam is added to that of the fly-wheel, and their combined force is exerted directly in splitting the cylinder, or tearing off the cylinder cover, thence effectually demolishing all the rods and gearing.

Where a steam engine, or rather the boiler, is overworked, and the load on the engine not very regular, experiments tending to the foregoing effect may be said to be going on almost daily; and in many cases even without the owner being aware of anything of the kind, until his engine has been broken down a few times, and after the expenditure of a few hundred pounds, when, as it is expressively said in Lancashire, " he finds it out."

This tendency of the water to rise into the cylinder is always considerably promoted by the very usual situation of the steam induction-pipe at the back end of the boiler, which seems to arise partly from the constant circulation of the water, which causes a current at the surface to set in the direction of the length of the boiler from the front end to the back. This circulation of water takes place in all oblong boilers, with a certain velocity depending on the ratio that the intensity of the heat in the furnace bears to the quantity of water to be kept heated, and is entirely independent of the other two causes before stated. But it is most probable that all three are sometimes combined in producing *waves*, which take their rise over the fire, and gradually increase in height as they pass towards the back part of the boiler.

That *waves* are always generated within a long boiler, when the engine is about to prime, is a singular but well-ascertained fact, as is shown by the frequent great and sudden depressions of the float at such times, especially if the latter happens to be placed at the contrary end of the boiler to that where the steam pipe is fixed. In watching the rapidly-successive alternate elevations and depressions of the indicator, buoy, or float of a boiler in this condition, the priming may frequently be observed to recur periodically after intervals of a certain number of strokes, provided that the state of the fire and the load on the engine continue perfectly uniform.

Those who have leisure and inclination to pursue this part of the subject, will find it an interesting field of inquiry, and far from being one of barren curiosity only, inasmuch as a few well-directed experiments may lead to results, perhaps, calculated to unravel some of the still unexplained causes connected with violent steam-boiler explosions.

SECTION 18.—ON BOILER ROOM GENERALLY.

From what has appeared, it would be fair to assume that the space for steam in a boiler *ought not to be less* than about half a cubic yard per horse power ; and when I have succeeded in establishing this position in the minds of my readers, by showing how universally this proportion is identified with the most successful practice, there will be no difficulty whatever in obtaining their assent to another proposition, namely, that the space required for water certainly *ought not to be more* than half a cubic yard per horse power. Therefore, generally, the total capacity of steam boilers may be stated at a cubic yard per horse power, not more than half of that space being water room, nor less than half of it for steam.

In enunciating this proposition I disclaim any right to being it first promulgator, for it used to be a current saying amongst the engineers in the north of England more than thirty years ago, that a good steam engine ought to have an area of piston equal to 27 circular inches for each horse power, while the boiler ought to have a capacity of 27 cubic feet for the same. This is one of those maxims which soon become established in any profession where there is a good deal of practice, and it is about as old as the time of the first introduction of Boulton and Watt's patent engines to the collieries in the counties of Durham and Northumberland.

In so wide a field for experience in the application of steam power to an *uniform kind of work*, the above maxim may be supposed to have some *à priori* claims to be considered somewhere near the truth. I have, however, endeavoured to prevent such considerations from influencing my opinions without further proof; besides, my experiments and observations have been for more than twenty years principally confined to the manufacturing districts of which Manchester is the centre, and more especially to the engines and boilers of the cotton factories, where also, from a *uniformity of work*, we have a similar field for experience to that presented by the collieries in the north of England, but with considerably greater facilities for investigation and comparison. The concentration of the cotton-spinning business in large towns, where the constant interchange of opinion amongst the managers and engineers of the various factories tends to produce a certain degree of uniformity of management, whilst the rivalry of competition will not allow any one to lag very far behind the best of his neighbours, has given decided advantages in this respect to Lancashire.

In stating the quantity of boiler room necessary for each horse power, there is no need to insist that it does not admit of considerable latitude, in which the difference in effect is scarcely perceivable; but the correct proportion is much easier ascertained than persons unused to such researches might expect.

It may surprise some engineers, who have been accustomed to calculate to the fraction of a horse power, to be told that the power of the steam engine itself can, in general, only be approximated to in a very rough way, compared to which the properly-calculated power of the boiler, with certain given qualities of fuel, is exactness itself. This arises from the friction of an engine, which, although capable of being ascertained within moderate limits for any assigned moment, is of that fluctuating character, and withal so large a portion—from *one-eighth* to *one-third* of the whole load—that there is nothing about the data of a boiler to be compared to it.

It is also much easier to acquire a knowledge of a number
of experimental truths on this subject, by a very little observation and attention, than is usually supposed, and without
directly instituting experiments, or wasting time by what is
very often misunderstood to be *experimenting*, that is, "*playing*
with models."

As a description of the best mode of experimenting on
a large and practical scale, we may suppose, what is a very
common case, that any given steam engine has two boilers,
with which it is worked alternately, each having the same
evaporating power, and in all other respects similar except as
to capacity, one of them having 20 and the other 25 cubic feet
per horse power (reckoned on the actual evaporation of the
boiler in cubic feet), of course a direct reference to the coal
account will give the comparative economy of the two boilers,
which we will suppose to be determined in favour of the larger
one.   Again, suppose we find, in another similar case, taken
at random, that the proportional capacities are 23 and 27
cubic feet per horse power, and that it is again determined
that the larger boiler works the engine with the least quantity
of coal, we are then assured that the true proportion is *above
twenty-five cubic feet* per horse.   Now if we take two other
sets of experiments on boilers, whose capacities we suppose to
be considerably above the true proportion, say 30 or 40, or
any other number of cubic feet per horse, and compare them
in a similar way, we soon arrive at the fact that the best proportion lays between 25 and 29 cubic feet ; and by proceeding
in this manner as opportunities for observation occur, we may
approximate to the true proportion as near as ever we like, in
fact to about *twenty-seven cubic feet ;* for the illustration given
above accurately describes the mode I actually adopted for
arriving at this result.

Although the above comparatively large proportion of boiler
room is given as the average of the best practice in Lancashire,
it must be stated that it is not at all a prevailing opinion
there that so large a proportion is needed.   The more popular

rule in Lancashire is, "for each square inch in the area of the piston to allow a cubic foot in the boiler." Now if the cylinders are of the ordinary proportions, it will be seen that this is little more than ¾ths of what we have concluded on as the best, which is more nearly at the rate of a cubic foot to a *circular* inch; and in the north-country rule before mentioned it is exactly so.

Those who have acted upon this popular rule, however, have been principally the makers of small engines, that are used for a variety of purposes where economy of space is of more importance than economy of fuel, and where, for the same reason, the engines are made with very short strokes. It has been observed, also, that this popular rule has given a proportion that has worked well where engines are working in pairs and not much overloaded, similar to marine engines; and from the fact that both Boulton, Watt, and Co., and some other of our best engine makers, have, during the last ten or fifteen years, erected several of those "land marine" engines in the cotton factories, the rule deserves some consideration.

It is, however, necessary to bear in mind that the object of a treatise like the present, professing to lay down a sound foundation for practical boiler engineering, ought to be first simply to determine the best proportions of the *single* boiler of a *single* double-powered condensing engine, and without complicating the subject more than necessary until the data for this one object is fairly established.

SECTION 19.—THE "BOULTON AND WATT" BOILER.

It is very commonly stated that Boulton and Watt allowed 25 cubic feet of space in the boilers for each horse power; but it is certain that Mr. Watt never left any opinion to that effect on record, and it is no less certain that the practice of the late firm of Boulton, Watt, and Co., gave no sanction to such statement.

In thus referring to the celebrated firm of Boulton and Watt, I do not profess to have any exclusive knowledge of the rules they used otherwise than by the means that are open to others, that is, by measuring a good number of the boilers and knowing their capabilities. The proportions used by other engineers I also preferred to obtain in the same way, although kindly offered every information on the subject by some of the first houses in the trade.

It is true that some of Boulton and Watt's 20-horse boilers that I have met with just occupy a space of about $18\frac{1}{2}$ cubic yards, or measuring by their extreme dimensions, say 12 feet long, $5\frac{1}{2}$ feet wide, and $7\frac{1}{2}$ feet deep—nearly 500 cubic feet—which, divided by 20, gives just 25 cubic feet of "space" for each horse power. But this is of course *space for stowage*, or room for the boiler to stand in, and not *inside* "boiler room," as usually understood, although it is not at all improbable that the general currency of the statement may have had no better origin than a mere commercial arrangement for packing it on board ship.

There is only one other way in which such an erroneous statement could have obtained currency, and that is by Boulton and Watt's 30-horse boilers being mistaken for 20-horse, owing to their ordinary practice, when erecting their engines in Lancashire, of putting down a 30-horse boiler to a 20-horse engine.

Although it is admitted that the day for wagon boilers is past, yet it may be useful to put on record the details of some of the most efficient and economical that are now in use in the Manchester factories, if it be only to mark the progress that may be made with circular boilers and greater expansion.

The following (fig. 5) is a cross section of a "Boulton and Watt boiler," as erected by Boulton, Watt, and Co., at a silk factory in Manchester, about fifteen years ago; and as this was a case (and not an isolated one) in which this particular kind of boiler was adopted by my advice, I have had special

Fig. 5.

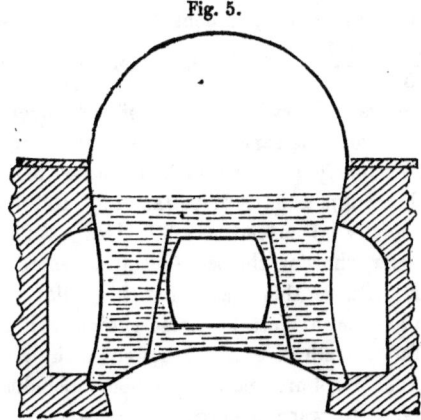

opportunities of watching its performance.   Fig. 6 is a longitudinal section of the same boiler, showing the uptake of the
inside flue.

Fig. 6.

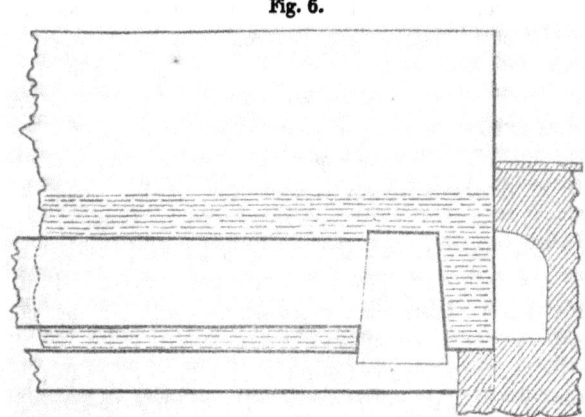

This boiler is 5½ feet wide by 7½ feet deep, the same as the
20-horse boiler mentioned above; but instead of 12 feet, it is
15 feet long, and contains an inside flue measuring 20 inches

wide across the top. This, according to the rule for flued boilers formerly mentioned, is considered in Lancashire equal to $\dfrac{(5\cdot5 + 1\cdot66) \times 15}{5} = 21\cdot49$, or about $21\frac{1}{2}$-horse power; and if we calculate the *cubic capacity* of the boiler, including the flue tube, from the same dimensions, we shall find it to be about 18¾ cubic yards, or very little more than 500 cubic feet; and this, supposing it to be taken for a 20-horse boiler (which it actually is) instead of a 30, gives just 25 cubic feet per horse power, showing the same result as before.

I shall only add in this place, that this boiler was equal to drive a Boulton and a Watt 20-horse engine with great ease, loaded to full 30 effective horse power, with less than 10 pounds of coal per horse power per hour, including making sufficient steam to warm a large factory where it is yet at work.

With respect to the effect which the propagation of the above assumed rule for boiler room has had on the practice of boiler making generally, there is no doubt but it has been beneficial; for makers and users of boilers have been naturally induced not to depart very far from what they considered Boulton and Watt's standard. We have, in consequence, a great number of boilers of all the various forms in use, ranging within a few feet above and below this proportion of 25 per horse power, from which a good average proportion can be obtained far more nearly correct than if the practice of engineers had varied at random, or to a greater extent on either side of this imaginary standard, which accidentally turns out to be very near the correct proportion. It is a comparison of a great many cases of this kind which has convinced me that those boilers which are a little above 25 cubic feet per horse are the most economical. It may be asked how we are to reconcile the above conclusion with the generally-acknowledged efficiency of Boulton and Watt's boilers, seeing they are not more than two-thirds the usual proportion of capacity given by other engineers, and less than two-thirds of what I now recommend as giving the best results in practice. This question admits but of being answered in one way, that is, by a reference to the practice already alluded to, of rating the boilers at *one-half more* than the nominal power of the engines they are intended to drive. The propriety of the practice was proved in the particular case recited, the engine to which this boiler was applied being overloaded just in the same proportion.

SECTION 20.—CAPACITY OF WATER CHAMBER.

In fixing on the proper capacity of the water chamber of a steam-engine boiler, there are not such peculiar difficulties as in the case of the steam chamber; and any one at a first view of the matter would say, as many do say without sufficient consideration, that there cannot be too little water, provided the boiler is filled to the proper height; for it is quite obvious the smaller the quantity of water, the less will be the expenditure of the fuel during the first getting up of the steam after each stoppage of the engine. It is, however, not the "getting up" the steam, but the *keeping it up*, that ought to be considered of most consequence.

It is a prevailing opinion that, after the steam is once got up, there is no material difference between keeping a large quantity of water boiling, and a smaller quantity, provided the escape of heat is prevented by sufficiently clothing the boiler with non-conducting substances; but on this subject engineers differ, although why practical men should differ in opinion on so plain a matter is unaccountable. It appears very clear to me that a *large quantity of water* must require more heat, or *heated surface*, to *keep it boiling*, than a smaller quantity, even supposing the heat required to generate the steam to be equal in each case; for there must be a great deal of power expended in keeping the water in motion, and every practical mechanic knows that we never get power for nothing.

On the other hand, when there is too small a quantity of water, it is difficult to keep the steam sufficiently steady. It is then quickly got up, but is liable to get quickly down again. This is more especially the case where the old system of firing and stoking by hand is still in use; but where any system of machine firing is used, on the principle of continuous supply, a much less quantity of water is found to do than was formerly thought necessary.

The priming of the engine is, also, not altogether unaffected by the quantity of water the boiler may contain, irrespective of the height of the water surface, inasmuch as a smaller quantity of water becomes much sooner *thickened*. The daily accumulations of whatever dirt or impurities enter the boiler along with the supply water, either in solution or suspension, become sooner concentrated by boiling; consequently frequent cleaning of the boiler, by preventing priming, enables us to work with a smaller quantity of water. It is in this way that there is almost always a saving of fuel to be effected by a proper arrangement of deposit vessels in the water chamber of a boiler; for they displace a certain quantity of water, and at the same time collect the deposit. Except for considerations of this kind, it is very evident that we cannot have too little water room; but our business at present is only with the ordinarily well-managed boiler of a factory engine, that is supplied with good water and kept moderately clean; and with that I consider half a cubic yard of water room per horse power ought never to be exceeded.

It has been very commonly considered that 10 or 12 cubic feet of water per horse power is as little as ought to be allowed. Tredgold recommends not less than 10, in consideration of the feeding apparatus for water not acting with perfect uniformity, even if ever so delicately adjusted; but in fact, contrary to what Tredgold supposes, it is usually found that the feed-water enters the boiler with the greatest uniformity when the feeding apparatus is not so *very* delicately adjusted; for when the float acts rather stiffly, it is then not so much affected with the occasional ebullition and agitation of the water.

I have already stated that we ought in no case to have *more* than 13½ cubic feet, or half a cubic yard of water per horse power: and lowering the water surface in a boiler at any time would go far towards convincing any one of this, were it not that the improvement thereby made may be supposed to be as much owing to increasing the steam room as diminishing the space for water. To obviate this source of uncertainty, however, I undertook a long course of experiments in partially filling up the water spaces of boilers with large stones and other articles without altering the water level or the other conditions of the boiler, somewhat similar to what is represented in figs. 7, 8, and 9, to the extent, in some cases, of leaving only a mere shell of water between the solid filling-up blocks and the boiler sides; and it was invariably found that

Fig. 7.

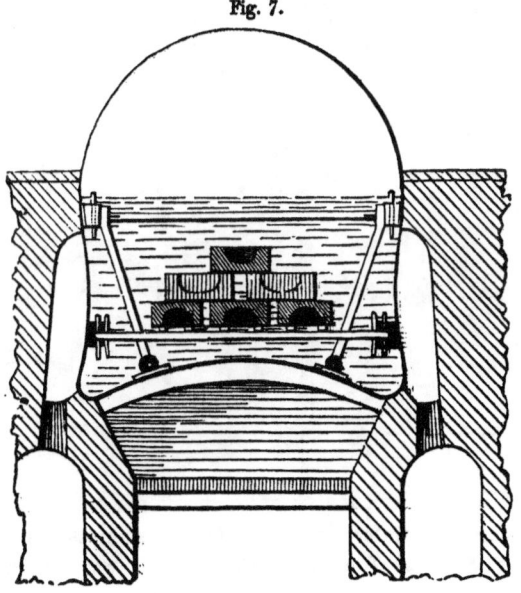

Fig. 8.

Fig. 9.

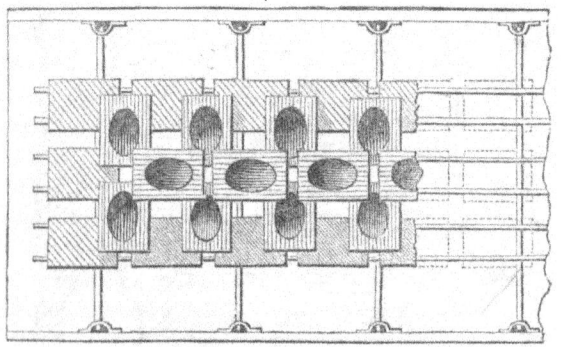

decreasing the water decreased the consumption of fuel.
Even as much as from 5 to 10 per cent. was occasionally

gained in this way. When I first mentioned the result of some of those experiments, the plan of using filling-up blocks in the water room was immediately adopted at several works in the north of England, although often without judgment or discrimination, and where it did more harm in one respect than good in another. Nothing could be more simple and cheap, literally costing nothing; for anything might be used for the purpose that came to hand, such as iron, stone, timber, &c. The articles used should be fixed upon a platform made of iron rods, laid across the internal stays of the boiler, and sufficiently clear from the bottom and sides, as represented in the preceding figures. The most convenient articles for the purpose were large fire bricks, from 18 to 20 inches long, 12 inches broad, and from 4 to 6 inches thick, with an oval-shaped concavity on one or both sides, sufficient to take away about one-half of the material. With these dimensions, 36 of them occupy about a cubic yard.

It is proper to mention one practical disadvantage in using the fire-brick blocks to any great extent. They were found to keep the steam up for some time after the fire was put out; a fact decisive of their economy in a theoretical point of view, yet, owing to the waste of water caused by the steam blowing away at the safety valve during the night, or after the engine had stopped, there was danger of the water getting too low.

There can be little doubt that these filling-up blocks have a beneficial effect in producing a uniformity in the evaporation, independently of the effect arising from the displacement of the water, and which may be likened to that of the fly-wheel on the engine, by becoming a sort of reservoir of heat, or regulator of the evaporating power of the boiler. This effect is especially observable if compared to the irregular action of a boiler with a large inside flue when fired by hand, the steam being liable to vary very considerably every time the furnace door is opened.

The plan which most readily suggests itself for diminishing the quantity of water in land boilers, without diminishing its depth or lowering the water level, is that of putting one or more flue tubes through the lower part of the boiler. It is the most usual course recommended by boiler makers; and, when not too expensive, is certainly the best, as it both adds to the heating surface and strengthens the boiler at the same time, the tubes being made to answer all the purposes of stays. When the boiler is too short in proportion to its diameter, it may also be serviceable, if required, in causing the current of flame and smoke to continue longer in contact with the boiler, which leads to the consideration of the subject of the next chapter.

CHAPTER IV.

Principles determining the proper Length of Boilers, with Examples, Rules, and Instruction in setting up.

SECTION 21.—ON GREAT EXTENSION OF SURFACE, AND SAVING WASTE HEAT.

ONE of the most prevalent sources of error, to scientific not less than to practical men, is the apparent paradoxical fact that, whatever length a boiler is made, the heated air or smoke which escapes is still capable of boiling water in a separate vessel placed in the flue leading to the chimney. This naturally induces a suspicion that, in our ordinary methods of setting boilers, we only obtain a small portion of the heat derivable from the fuel. Some of our most eminent scientific men have publicly stated as their opinion, derived from "chemical considerations" only, that, owing to "imperfect combustion" alone, to say nothing of the misapplication of its products, some *sixty* to *seventy* per cent. of the fuel is wasted! And nothing is more common than to hear the most *practical* men declare their belief that, for want of a proper application

of those products, "one-half of the fuel goes up the chimney."
Hence our patent offices are filled with the thousand and one
schemes, with their various and endless winding and zigzag
flues, of the numerous inventors of boilers, many of whom
appear to me to have quite a mania for running after and
"using up" the "whole of the heat," with a determination
far exceeding that of the perpetual motion seekers, and with
quite as little chance of success.

The fact stated, however, is in no way surprising, although
steam may be thus raised in a close vessel, even to a much
greater pressure and temperature than that in the boiler from
which the waste heat has escaped. It certainly must be
allowed that steam so obtained, if returned into the boiler at
a workable pressure, is *so much* clear gain. My argument,
however, only is that it is *not much;* and calculated commer-
cially it is worth less than nothing, that is, taking *time* into
the account; for, according to what is elsewhere observed, we
find the steam obtainable in that way to be produced at so
slow a rate, that its value is less than a very small per centage
on the capital employed to obtain it.

The general leaning of many manufacturers, as well as the
great majority of engineers, to erroneous opinions respecting
the proper length of boilers, as well as the proper quantity of
heating surface, namely, that they may both be almost inde-
finitely extended with advantage, is much to be deplored, and
is difficult to eradicate, except by dear-bought experience—the
best and only proper teacher where parties, from prejudice,
interest, or something worse, are inaccessible to reason. The
only misfortune is that, of the two classes just referred to,
manufacturers or proprietors of steam engines and engineers,
the dear-bought experience of the latter is *commonly paid for
by the former !* This needs no comment as to the result to be
expected. I can only say that, both previously to and since
the publication of the last edition of my practical Essay on
Boilers, in 1839, I have had many opportunities of witnessing
the effect produced by lengthening boilers, but never yet met

with a single case where any absolute saving accrued from such alterations, excepting where there had been previously a manifest deficiency of heating surface or of steam room.

The exceptions just stated are important, and require to be particularly noticed, for the neglect of them is generally the cause of much self-deception on this subject. A manufacturer, for instance, has got his engine overloaded, causing an increased consumption of fuel, arising from the necessity of overfiring a boiler too small for the work it has to do. He then finds, perhaps for the first time, that the temperature inside the chimney flue is 500° or 600°, and knowing that the heat of the steam is little more than 212°, he hastily concludes that the difference is sent up the chimney and *lost ;* and consequently, instead of putting in a larger fire-grate, or at the most a larger boiler, he gives a ready assent to the use of any plausible means of "*robbing* the smoke of this waste heat before it escapes." Hence a boundless field of speculative inquiry is opened, and how to "use up the whole of the heat" in the most advantageous manner, becomes a question of intense interest to those whose establishments require the expenditure of several thousands a year in fuel.

In many attempts to save this waste heat separate boilers have been added, completely inclosed in the main flues leading from the boiler to the chimney, without the least perceptible result as to saving of fuel. These supplementary boilers, or "heat savers," as they are always called (when first ordered), have been also made without any steam room, and applied solely for the purpose of heating, or rather only warming, the water for feeding the larger boiler. It is true that the water is certainly elevated in temperature on its way to the boiler; yet exactly how much heat has ever been saved by such means I believe is yet a question for the philosophers; but judging of the quantity of heat, latent and sensible, saved by its effect in saving fuel, which is the most sensible criterion for the manufacturer to judge by, it differs from nothing (if not quite nothing) by less than any assignable quantity.

In one case to which my attention was particularly directed at the time, from the circumstance of its being adopted, if not at the express advice, at least in accordance with the opinions, of a very eminent Manchester chemist, the "heat saver" was a 4-horse cylinder boiler, of 2 feet diameter and 10 feet long, suspended longitudinally in the main chimney flue of a 40-horse Boulton and Watt boiler, in such a way as to be no interruption to the draft; and although this might be considered equivalent to the addition of at least 10 per cent. to the heating surface of the boiler, in order to make the experiment more decisive a second "heat saver" was at the same time ordered, of somewhat thinner iron, but of exactly the same size, and also fixed at a little distance beyond the first. They were connected together by a pipe, and the first was connected with the ordinary feed pipe of the boiler. A separate supply pipe was also carried from the hot well to the farthest heat saver, so that the feed water was forced through both the heat savers before it entered the boiler. The ordinary feed apparatus was also retained, so that either mode of supply could be used for the purpose of experiment. The chief result, however, was clearly established to be no appreciable saving in fuel; and although the heat savers succeeded in making the water within them nearly, if not quite, as hot as that in the boiler, yet it was observed that there was very little difference in the temperature (only two or three degrees) within the ordinary feed pipe of the boiler, whether it received its supply from the heat savers or direct from the hot well of the engine, at least 100° colder.

This result was evidently a consequence of the fact, that the quantity of heat required to raise the temperature of the water in the boiler, about 100°, was practically nothing as compared to the quantity required to convert the same water into steam.

Section 22.—Great Extension in Length not necessary.

The plan most commonly resorted to, however, is to lengthen the boiler, and if in any such case a manufacturer finds a certain saving of fuel, he usually ascribes the improvement solely to the extension of the distance the smoke and hot air have to travel before they escape, forgetting that the increased quantity of heating surface and boiler room may be quite sufficient to produce the increased economy in fuel, which in fact is the case nine times out of ten.

Correctly recorded experiments in all cases support the position I wish to enforce; which is, that in a well-proportioned boiler *there ought always to be a sufficient area of heating surface within as short a space as possible.*

The propriety of a practice founded upon this principle was conclusively settled at that great era of steam engineering, the opening of the Liverpool and Manchester Railway, by the successful trial and final adoption of Mr. Stephenson's great invention, without which railways would have been useless for many of their most important purposes—I allude to the firebox tubular boiler of the celebrated "Rocket" locomotive engine: to say nothing of the engine itself, as being a matter of mere secondary importance for connecting the power with the wheels,—it was the *Boiler* that was pronounced, with scarcely a dissentient among the hundreds of engineers who witnessed its performances, to be "*the* invention" that was to "make" the railway; a prediction that a triumphant career of twenty years without a single competitor deserving the name has amply verified. It has in truth made railways, not only figuratively, but actually.

The boiler of the Rocket, which so completely transformed the slow-going "puffing billy" travelling engine of the Hetton "wagon-way" into the iron race-horse of Lancashire, at the same time added greatly to its power of doing heavy work. And the more closely the principles involved in that great invention have been adhered to since it first " burst on the world like a rocket," the more perfect has been the railway locomotive for efficiency and economy combined ; although the most valuable of those principles were either not very clearly understood, or not candidly acknowledged, by any party during the celebrated battle of the gauges. And as it is always good to recur to first principles, it may be as well to remind railway engineers that the great principle of the locomotive boiler is *quick* combustion wilh *short* and *direct draft*, the extreme opposite of the far-famed Cornish system of slow combustion with long drawn out flue surface, and points clearly to the policy of increasing the number of tubes, and consequently the diameter of the boiler rather than its length.

In adverting to the advent of the first efficient locomotive, it is impossible to help recurring to the strenuous efforts then made, in Liverpool and other places, to prove by experiment the alleged superiority and economy of various new patent boilers, constructed on a principle completely opposed to that of the Rocket,—having long continuous winding flues. In one of the patent boilers it was alleged that " the heat was all so entirely used up," by its extraordinary economical qualities —that the caloric evolved from the fuel was so completely abstracted by the rapid generation of steam, and the heated air and incombustible gases so far robbed of their heat and cooled down, that *the naked hand or arm might be placed with impunity inside the tube* by which the smoke escaped from the boiler. And as if this experimental specimen of proving too much was not sufficient, it was gravely alleged to save *one hundred and twenty* per cent. in fuel ! !

SECTION 23.—PRINCIPLES GOVERNING THE LENGTH OF BOILERS.

When a boiler is to be set up in the simplest possible way, that is, without return flues of any kind, it is of some consequence to know to what length the flame is likely to extend, because on this will depend in a great measure the extreme limit of the length of the boiler.

In using fuel with little or no flame, it is evident that a boiler of a form approaching as nearly as convenient to that of the equal cylinder, or cylinder of greatest capacity, would be the best for economy in *first cost,* and if made with hemispherical ends, would also be the best for *strength*. Therefore a boiler of this form, whose length is equal to twice its diameter, will be the *shortest* that can with any degree of propriety be recommended in any case. And it is possible that such a boiler might be beneficially adopted in some cases where coke or anthracite coal is the fuel used.

There is a current opinion amongst experienced workmen, that the common wagon boiler ought to be about three or four times the length of the fire grate; and it is based upon the observation, that when a boiler is set up in that proportion, and the fire "managed as it *ought* to be," that the whole of the flame will be expended against the boiler bottom, and never pass into the side flues along with the smoke. The truth of this observation, however, depends upon circumstances. If heavy charges of some of the more bituminous kinds of Newcastle coal are made upon a grate with thick bars and confined air spaces, it is far too common to see the flame extend occasionally to double the length of the boiler every time it is roused up by a very determined stoker; and it is commonly enough done as a feat to be boasted of; it is, however a feat of barbarism that, except in some possible urgent circumstances, ought never to be attempted.

With boilers whose fire grates are square and whose lengths are not less than four times the length of the fire grate, I have never seen an instance of the flame reaching the end of the boiler, where I have not at the same time found either the fire bars too thick or the fire too thick upon the bars, or a bad draft, and usually all three together. Even when the chimney draft is good, should the furnace not be kept regularly cleaned and the fire bars free from clinkers, or should there be a large mass of coals thrown upon the fire at once, covering the whole of the area, and thus interrupting the due supply of air through the grate, the consequence is nearly similar. In such a case the furnace becomes little better than a gas retort, and the gas so made has been frequently known to fill the chimney and flues, so as to take fire at the chimney-top, and burn like a gas light. This thick firing is in fact the ordinary trick of smoke burners. "Distilling" gas, as they term it, from heavy charges of coal, and then setting fire to it beyond the bridge, thus occasionally filling the flues with flame, but generally with a feebly burning mixture only of carbonic oxide gas and air, whilst in the furnace itself, for want of a due supply of air through the grate, there may be very little heat or flame produced. These, however, are exceptional cases ; but where the ordinary *open firing*, instead of *close firing* or "charging," as the destructive system of *heavy firing* is very properly, though innocently, called by its advocates, is used, and there is a free supply of air (where only it ought to be) through the fire grate, the flame from common coal, if well spread out in a thin sheet against the boiler bottom by means of properly-constructed bridges, will ordinarily be expended in about twice the length of the furnace.

In accordance with these and other reasons before given, I have always preferred having as much of the boiler bottom as possible exposed in the furnace to the direct action of the fire and flame.

The proper area of heating surface to the fire grate being now determined upon, with a proportional space for the flame to develope itself in, the question next to be decided is, whether it is better to obtain that surface in as compact a space as possible, in the immediate vicinity of the fire grate, or by means of comparatively long narrow flues, either winding or otherwise. That the first is infinitely to be preferred to the latter arrangement is so obvious, that it appears strange how a contrary practice could have crept into use.

As a general principle, then, in the construction of boilers, we must select that form which gives the proper quantity of heating surface along with the greatest capacity as well as strength for the quantity of material employed, and withal comprised in as compact a space about the fire-place as possible. The three last conditions would lead us to prefer the spherical form, and if it were absolutely necessary to have the furnace within the boiler, a form approaching to this might still be the best, as in the fire boxes of some locomotive engines. The old-fashioned "haystack" boiler is also a near approach to this shape, and up to a certain size it is yet, perhaps, as good as any when well made.

The next in degree for simplicity of form is the horizontal cylinder, and it is almost a necessary requisite, to ensure the greatest strength and perfection of workmanship, that it be made without any inside flue. This form of boiler admits of a fire grate underneath it of its full width; and, if set up without external side flues, and by means of proper bridges, the flame may be made to envelope nearly the whole of the lower half of its external surface very advantageously. There is, however, an admitted difficulty in adapting the fire upon a level grate, so as to act uniformly against the convex boiler bottom; but their greatest faults have arisen from imperfect methods of setting, being commonly set up with side flues, like wagon boilers, to which, unless of very large diameter, their form is not at all adapted.

SECTION 24.—EXPERIMENTAL PROOFS IN FAVOUR OF SHORT BOILERS.

Whether there is anything to be gained by making a boiler of greater length than four times the length of the fire grate, other conditions being the same, there have been very few experiments on a large scale to determine. The only one of which the result has been published, bearing in some degree upon the question, is one by Mr. Stephenson, of which some account is given in Mr. Nicholas Wood's work on Railroads.

The experiment was made with a boiler similar to that of the Rocket locomotive, when the fact of the very much greater evaporating power of the fire-box, per unit of surface, than of the rest of the boiler containing the tubes, was decidedly proved; and the difference found to be so very considerable, about 3 to 1, that it had great influence in producing the present practice of making locomotives with very large fire boxes, and which has proved to be so eminently successful.

M. Pambour regrets that such experiments had not been more frequently made, and without which he appears to think it a difficult matter to settle the true theory of the locomotive engine. This was not surprising a dozen years ago, when M. Pambour's own disinterested and valuable experiments were made on the Liverpool and Manchester Railway. But important as such additional experiments would be, I believe they are still a desideratum so far as the locomotive is concerned. This is perhaps neither the century nor the country, for carrying out such experiments, unless some direct practical benefit resulted to some one monopolist or monopolising company. The proper length of a boiler is not *patentable*, otherwise it would have been found out long ago. The fact is, that private individuals cannot undertake the expense of experiments on a large scale, and many shrink from the labour.

Even with respect to the ordinary factory boiler, it is no very easy matter to induce manufacturers to try experiments of any kind. Although I have for many years urged on them the importance of settling this particular subject of length, when about to order new boilers, I have never succeeded, even partially, except with two or three who took some few steps towards it. I never made but one direct experiment to this end, which was necessarily on a small scale, of which I published an account at the time. It was as follows :—I made an open-topped boiler divided by water-tight partitions transversely into four equal divisions, each nearly filled with an equal quantity of water ; a coal fire was made upon a square fire grate, underneath, and close to one end, of exactly *one-sixth* the length of the boiler; the coal was supplied in the usual manner by hand, and each division of the boiler was similarly supplied with water. It was then found that the proportional quantities of water evaporated in the first and second divisions were something like those in Mr. Stephenson's experiment; but in the third division the evaporation was still more greatly diminished; and in the fourth it was practically nothing, or so small as scarcely to bear any comparison to the whole.

The result of this experiment was sufficient to determine— what it has since been my uniform practice to recommend— that the length of boilers need not in any case exceed *six* times their width or diameter; supposing the latter to determine the size of the square fire grate. This advice has been extensively acted on for many years with uniform success. In one instance, where some plain cylindrical boilers, equal to six and seven diameters in length, were found to be about as economical as any other kind previously used, other boilers have been made for the same works successively, equal to *five* and *four* diameters, with equal if not increasing economy in fuel. In this instance I am of course speaking of boilers with direct draft, and no internal flues nor flues of any kind.

When a cylindrical boiler is large enough in diameter to allow of its being set up with winding flues or a wheel draft, of course its length may then be advantageously reduced to the same proportion, or nearly, as in wagon boilers that are set up in the same manner, or to about *four* times their diameter.

This, however, is a method of setting cylindrical boilers which is decidedly objectionable. For unless the boiler is 5 to 6 feet in diameter, or upwards, the seating walls in such a case generally encroach too much upon the necessary width of the fire grate; or, if the furnace and fire grate be made sufficiently wide, then the side flues will be found too small for the draft or for convenient cleaning, unless the brickwork is gathered in above the central line of the boiler, thus making the side surface almost useless as effective heating surface.

Besides, this small space for the side flues often induces the bricklayer to cover in the flues nearer to the water level than is prudent, and then it is pregnant with danger from the greater liability of the water occasionally getting too low.

SECTION 25.—PROPORTIONING LENGTH TO QUALITY OF COAL.

When a cylindrical boiler has an inside flue tube, and is set up with a split draft, as described in sec. 5 (p. 9), then the above objections do not so fully apply ; because the smoke being then divided into two currents, the side flues only require to be half the area of the former. In this case, on account of the inside tube, the boiler cannot be conveniently less than 5 feet in diameter, and it need not be more than about $3\frac{1}{4}$ diameters in length. Thus a boiler of this diameter may be $17\frac{1}{2}$, say 18, feet long, and, with an inside flue tube of 2 feet diameter, will have about 22 square yards of effective heating surface, and with 22 square feet of fire grate, say 4 feet wide by $5\frac{1}{2}$ long, will be a 22-horse boiler ; as appears by the following example of the calculation, according to the formula for the short-slide rule (at page 13) : only instead of using Mr. Hick's divisor of $5\frac{1}{3}$, use 5·73 for a gauge point, as recommended at page 23, which gives the number of square yards of effective heating surface, thus :—

A	Gauge Point = 5·73	$d = 5 + 2 = 7$
ɔ	sq. yards = 22	Length = 18

The above is for Lancashire or any other inferior coal, requiring a square foot of grate bar per horse power. If a superior quality, say Newcastle coal, requiring only $\frac{3}{4}$ths of a square foot per horse be used, the power of the boiler will still remain the same ; but to prevent undue waste of heat it is then only necessary to reduce the length of the grate in the proportion of 4 to 3, or from $5\frac{1}{2}$ feet to little more than 4. It is, therefore, necessary to bear in mind that the rules given above for regulating the length of boilers relate exclusively to the practice with Lancashire coal, and low pressure, stationary, or factory engines, and that where better coal is used the grate need not be so wide, and the diameter of the boiler may be proportionally reduced, which would make it more suitable for high pressure. But in this particular case it is preferable that the boiler should be extended in length, rather than diminished in diameter; which would leave too little room for the flue tube, and alter the character of the boiler.

Taking the above example, for instance, and adapting the shape of the boiler to Newcastle coal, it must be lengthened one-third, thereby making it $4\frac{3}{4}$ diameters instead of $3\frac{1}{3}$ diameters long, say about 24 feet instead of 18 as before, the heating surface being of course increased exactly in the same proportion, or from 22 to 29·3 square yards : which last number is the number of nominal horses' power the boiler would now be equal to, if fired with Newcastle coal upon the first-mentioned grate of 22 square feet, or $\frac{3}{4}$ square foot per horse.

While making use of the above examples in illustrating the subject of lengthening boilers, it must be observed that any comparative experiment made with boilers of the precise dimensions here given would not be a fair one ; because, accurately speaking, the flue tube in the last example ought to be enlarged in its cross sectional area from 4 to $5\frac{1}{3}$ circular feet, or in proportion as the power of the boiler is increased ; that is, from 2 to 2·3 feet in diameter, supposing the same kind of coal used.   This would increase the power as follows :—

A	G. P. = 5·73	$d = 5 + 2\cdot3 = 7\cdot3$
ɔ	H. P. = $30\frac{1}{2}$	Length = 24

namely, to $30\frac{1}{2}$-horse power.   The enlarged flue tube would also alter the proportions of water and steam room, to provide for which a separate steam chamber should always be contrived.

When a flued boiler, either cylindrical or wagon-shaped, has the uptake inside the boiler, as in a Boulton and Watt boiler (fig. 6), which is an excellent arrangement, so that a portion of the flame may occasionally pass into the tube instead of being expended against the brickwork, the length of the boiler will then admit of being reduced to about *three* times its width or diameter. This, in fact, is the exact proportion of many factory boilers of both kinds, thus set up with a split draft, that have been found to be exceedingly economical.

SECTION 26.—CORNISH AND BUTTERLY BOILERS.

In Cornish and Butterly boilers, both of which have the fire to pass through the inside flue tube first, when this flue goes quite through to the end of the boiler, I have not found them to be quite so economical if made much less than *four diameters* in length : and this would seem to agree with the best practice in Cornwall, where, from long experience, exclusively confined to one kind of boilers, there can be no doubt that the maximum economical results have long ago been arrived at, and that is when the boilers are not more than *six diameters* long ; the average of about 30 boilers in two of the best-conducted mines in Cornwall being little more than 5½ diameters long.* Allowing for the superior quality of the Welsh coal used in Cornwall, which is probably from 20 to 40 per cent. better than the common coal used in Lancashire, this difference of one-third in the length of their boilers is fairly allowable, and is consistent with the best practice in both counties.

If, however, the fire flue tube of a Cornish or Butterly boiler have a connection with the boiler bottom without going through the end of the boiler, similar to the uptake of a Boulton and Watt boiler, which in this case may be called the

* I am glad to be enabled to make this statement from information kindly furnished me by one of the first authorities in Cornwall, and the more so from the circumstance of having inadvertently admitted into my former work, on what I now know to be insufficient authority, a statement to the effect that the average consumption of coal by *all* the engines in Cornwall was then about 6 ℔s. per horse per hour; an overstatement certainly, but not more so than has frequently been made by various public writers, advocates of the Cornish system, when referring to the comparative consumption of the Lancashire engines.

down-take, then the length of the boiler may be very advantageously reduced to *three-and-a-half diameters*. There are some Butterly boilers now working in Manchester of this kind, which I purposely designed in this proportion in preference to a greater length, being 8 feet diameter by 28 long, and they have never been exceeded in economy. These boilers were 9 feet deep, being oval in their transverse section; but there are many in that district of the same proportion, but circular, which is preferable for high pressure, say of 8½ feet diameter and 30 feet long, that are equally economical. Boilers in this proportion and of these dimensions, with a "take-down" inside, and set up with a split draft, if supplied with fuel by firing machines, and no stoking, are usually found to unite all the qualities of a good boiler that can be wished except one, and that is they make more smoke than those whose furnaces are more completely surrounded with brickwork, which is a difficulty that seems almost inseparable from all single furnaces in boilers that generate steam very rapidly, and is partly to be ascribed to the consequent rapid abstraction of heat from the fuel.

Excepting the wagon-shape Boulton and Watt boiler, which is inadmissible except for low pressures, the above-described Butterly boilers, when working to the greatest extent of their power, are certainly amongst the most economical boilers, taking both fuel and first cost, that are to be met with in the cotton factories. They are not so well calculated for very high pressure as the Cornish, but usefully occupy an intermediate position between them and the wagon-shape. They have been in very general use during the last ten years, and the greatest number of them have been made somewhat shorter in proportion than the above, which may therefore be taken as affording a safe guide in practice.

From the consideration that the above remarks, together with the observations heretofore made, point out the proper limits in practice for all kinds of stationary boilers as respects length, I have drawn up the following summary of the principles involved, in the shape of rules. These rules are not meant for the use of the boiler maker solely, whose best business often is to make a boiler to *fit* any place of any shape, but rather for the guidance of the factory architect in first setting out and arranging his designs for a building, in order to leave room enough for the engineer's plans, which architects very seldom do. This remark, it may be observed, might not inaptly be extended to some naval as well as factory architects, who undertake to design steam ships without bestowing a proper thought about the boilers.

SECTION 27.—GENERAL RULES FOR PROPORTIONING THE LENGTH OF BOILERS FOR STATIONARY ENGINES.

RULE 1.—A plain cylindrical boiler, without any inside flue tube, and hung on what is sometimes called the " oven plan," that is with a direct draft passing from the fire-place directly under the bottom of the boiler to the vent or chimney, and without return flues of any kind, need not exceed in length *six* times its diameter; and it ought not to exceed *six* times the square root of the area of the fire grate in feet if worked with Lancashire, Derbyshire, or Yorkshire coal, which is equivalent to *six* times the square root of the nominal horse power of the engine in feet. And if worked with the best Newcastle coal, the boiler need not be more than about 8 diameters long, and ought not to exceed 8 times the square root of the area of the fire grate in feet, which is equivalent to 8 times the square root of three-fourths of the nominal horse power of the engine in feet; and in any case it never ought to exceed *six* feet in diameter.

RULE 2.—A boiler without any inside flue tube, and set up in the common way, with external brick flues and a wheel draft, need not be more than about *four* diameters long, and ought not to exceed in length 4 times the square root of the area of the fire grate in feet for Lancashire coal. If worked with Newcastle coal, it ought not to be more than about

5 diameters long, and need not exceed in length 5 times the square root of the area of the fire grate in feet; and in either case, whether a wagon or a cylindrical boiler, it never ought to be more than 6, nor less than 4 feet in diameter.

RULE 3.—If a boiler contains one or more inside flue tubes, passing quite through it, and is to be set up with a split draft, it need not be more than about $3\frac{1}{4}$ diameters long, and ought not to be longer than $3\frac{1}{2}$ times the square root of the area of the fire grate in feet for Lancashire coal; and if worked with Newcastle coal, it need not be more than about $4\frac{1}{2}$ diameters long, and ought not to exceed $4\frac{1}{2}$ times the square root of the area of the fire grate in feet; and never be less than 5 feet in diameter.

RULE 4.—When a boiler contains one or more internal flue tubes, with an inside uptake or connection with the boiler bottom, as in Boulton and Watt's, or as in the marine "tubular" and other multiflue boilers, the length of the boiler need not be more than about *three* times its diameter with Lancashire coal, and ought not to exceed in length 3 times the square root of the area of the fire grate in feet; but with Newcastle coal its length in feet may be equal to 4 times the square root of the area of the grate, and need not be more than about 4 times the diameter of the boiler.

RULE 5.—Cornish boilers and Butterly boilers, set up in the best manner, with split draft, and using Lancashire coal, need not be more than about $3\frac{1}{2}$ and 4 diameters long respectively; and if using the best coal, they ought not to be more than $5\frac{1}{2}$ and 6 diameters long.

Boilers whose dimensions are proportioned within the limits stated in the above practical rules, more especially those indicated by the three last, are mostly the best to be found of each kind in the manufacturing and midland counties.

The manner of hanging boilers, to which the first rule relates, is yet generally confined to some country places, where inferior workmanship only can easily be obtained, and where the space occupied by the greater length of the boiler is not of much value; consequently the data for this rule have not perhaps been quite so exactly determined as may be. There are reasons for supposing that the direct-draft cylinder boiler may be made considerably shorter than the proportion of 6 to 1, as stated in Rule 1, if the fire could be equally well arranged as in the wagon boiler, the concave bottom of which is admirably adapted to this purpose, because the middle of the grate, where the heat is the most intense, is at the greatest distance below the boiler bottom, while the latter gradually approaches nearer to the grate at the sides of the furnace, and thus tends to equalise the action of the fire against the boiler. The bottom of the cylindrical boiler being convex downwards, the action of the fire is of course exactly the reverse of the above. To say nothing of the injury done to the boiler plates on this account, there requires to be a greater average distance between the grate and the boiler bottom; this again requires a greater quantity of coal in the furnace, which impedes the draft and renders stoking necessary, thus causing the flame to be occasionally extended in length; and although such undue length of flame cannot be kept up continuously with any degree of economy, it has given rise to a natural, though unfounded prejudice against this mode of setting a boiler, often expressed in the observation that "*all the heat goes up the chimney.*" Erroneous as the idea is that gives rise to this very common remark, it is not a little strengthened by the fact that the temperature of the chimney is always very much greater with a direct draft than it is where winding brick flues are used, which may be considered only as a portion of the chimney lying horizontally, the superfluous heat in which is doing no good, and is really "waste heat," from the great inconvenience and trouble it occasions; whereas when this waste heat is allowed to go freely up the vertical shaft of the chimney, it really becomes of great use in increasing the ascensional force of the current, thus improving the draft and enabling the boiler to be worked generally with the damper early closed, as all steam-engine boilers ought to be worked.

SECTION 28.—BOILERS ON THE OVEN PLAN LIABLE TO
EXPLOSION FROM SURCHARGED STEAM.

When a boiler is set up or "hung" with a direct draft, as
described in Rule 1, it is very commonly but erroneously said
to be on the "*oven plan*," which designation ought to be confined to those only where the flame is caused to pass wholly
or partially over the top of the boiler, although they have
generally a direct draft also. The use of such plans, however,
cannot be too much reprehended as pregnant with danger,
from surcharging the steam with heat, and thereby becoming
liable to explosion. Few, however, if any, are now set up in
that way; but I cannot help thinking that some of the otherwise unaccounted for explosions which have occurred of late
years have been owing to a similar cause.

Surcharged steam may be produced in a boiler without the
latter being exactly on the oven plan, and I have not the least
doubt that it is frequently, if not constantly, so produced in
cylindrical boilers that have the side flues carried up considerably above the level of the centre of the boiler.

Although out of place to treat on explosions in this chapter,
it is most important to omit no opportunity of inculcating
caution in setting up cylindrical boilers with or without side
flues, if the fire or flue is carried very high up or too near
the water level. It is more especially the case if the boiler
is of small diameter; the flame is then compelled to act
strongly against the top of the side flues, thus overheating
the brickwork, and reverberating partially *above* the convex
sides of the boiler, having only a slight effect in generating
steam.

Should, however, the water level happen accidentally to fall
only a single inch below the top of the flue, we know very
well how rapidly the steam would become surcharged with heat,
although, perhaps, no very great addition to the pressure would
be indicated by the steam gauge.

There is no doubt whatever, in my mind, that in such a

boiler, even without supposing the water to fall below its proper level, the steam is constantly being surcharged, owing to the boiler plates, in the upper part of the flues, being in such an unfavourable position for transmitting the heat to the water partially below them, at the place where the greatest heat is always found. There must be a particular liability to the accumulation of surcharged steam where the boilers are well covered with brickwork or other bad-conducting substances. The heat imparted to the iron is thus in a manner prevented escaping in any other way than by conduction up into the steam chamber and surcharging the steam. This process of heating the steam will go on at a slower rate certainly, but not less surely, than if the water had been for a short time too low.

Now, supposing this over-heating of the steam and the top of the boiler to occur from either of the above causes, while the engine is at work it is not perhaps likely that anything particular will be observed to ensue, unless, probably, the burning of the clothing or other covering of the boiler. But let us examine the matter when the engine has been standing for some time, *or when just about to start after the first getting up of the steam,* and we shall find a very different state of things, the consequences of which, if only leading to the slightest probability of resulting in an explosion, are too serious to be passed over without great consideration.

We may suppose then that the steam chamber has become filled with over-heated or surcharged steam from any cause whatever, whilst the top of the boiler is also in the condition already described, that is, exposed to a temperature of say from 350° to 400°; which supposition is quite consistent with the fact that the great bulk of the water in the boiler may be at the same time considerably below the common boiling point. And although this last assumed fact is, perhaps, only of rare occurrence in boilers which have their fires underneath them, it cannot be too widely known, that in all stationary boilers which have internal furnaces it is not only of frequent occurrence, but it is a common observation of most practical engineers and workmen, that at the first getting up of the steam from cold water in a boiler of this description, with the safety valve and all other outlets of the boiler closed, the *water at the lower part of the boiler is often quite cold*, whilst there is a pressure in the boiler of 10 or 20 lbs. on the square inch.

From what we know of the very slow conducting power of water for heat *downwards*, if some special means of providing for its due circulation be not used, it is quite conceivable that a comparatively *thin film* only of the surface of the water will correspond in temperature with the surcharged steam, in the case we have supposed, decreasing in some rapid ratio downwards to the level of the furnace. It is also evident that whilst the surface of the water remains perfectly still and undisturbed, the state of things may continue for some time, say for several minutes, while the super-heated condition of the upper part of the flues and the top of the boiler will go on increasing more or less rapidly, according to the extent of the flame that is kept up, communicating a correspondingly elevated temperature to the already surcharged steam, without adding very much to its quantity or pressure. The evaporation from the surface of the water will, at the same time, be also going on, slowly perhaps, but gradually increasing, at a rate which will then be in proportion to the superficial extent of the water surface, and the heat of the steam incumbent upon that surface.

As my object in thus tracing the proximate causes concerned in producing the dangerous consequences that may ensue is for the purpose of suggesting the means of prevention, it will be here useful to observe that, whatever may be the pressure of the steam due to the actual evaporation at this stage of the process, it will be nearly doubled by the heat of the steam chamber, supposing the latter to have arrived at about 400°, which is still under the temperature that would leave any permanent traces of its existence on the metal, the lowest temperature that gives the first discolouration to iron (a straw colour) being about 430°. Now, excepting by the thermometer, there is only one other means of giving the fireman any ready indication or suspicion of there being anything wrong about the boiler; and rough and uncertain as that indication is, it is always sufficient to create alarm, and induce him to take some instant precautionary step; and whether that step be properly or improperly taken will make all the difference between hastening and preventing the catastrophe of blowing up the boiler. The indication I am alluding to may be thus de-. scribed:—whenever a fireman finds that, on getting up the steam previous to starting an engine, more particularly after a boiler has been at rest and refilled with cold water, that the pressure rises to a certain height *in about half the time* that has usually been the case previously; or if, after firing a certain time, and consulting his steam gauge, he finds that the steam has run up to 10 or 20 lbs. pressure, instead of 5 or 10 lbs. as he expected, in the same time, and if the man has been accustomed to the same boiler, and is assured there is no unusual circumstances, of draft or otherwise, to account for the difference, then he may be almost certain that there is surcharged steam in the boiler.

Now an idle or a careless stoker might not perhaps notice the circumstance just described so long as the required pressure is not reached which the safety valve is loaded to blow off at; which is the case no doubt very frequently, and the circumstance is thus passed over without danger or the suspicion of danger. But what is the step that a *careful*, and still more a *timid*, man would be likely to take at such a time? In all probability, if the means were at hand, he would either "feel" or lift the safety valve; and what is the consequence that would be most likely to follow? In the case we have supposed it certainly would not be safety but danger, and, very possibly, immediate destruction; for, when a safety valve of any considerable area is thrown *suddenly* wide open, there is always a sudden rush of water, or rather water and steam mixed, to the opening. This is well known to occur universally under such circumstances, and I have more than once seen it purposely exhibited by foolhardy engineers, by way of *illustrating* the nature of priming, as *priming* in fact it is, at the safety valve instead of into the cylinder.

We need not, however, suppose that this is exactly the course taken in the case under consideration; for a *very* careful man would perhaps only ease the safety valve gently on its seat; and if it was in a cotton mill or other factory requiring to be heated by steam, the almost certain course he would take would be to let the steam into the pipes for that purpose; and whether he opens the communication suddenly or slowly, the effect produced is that the surface of the water is simultaneously relieved from a portion of the pressure, and in consequence of being so relieved the water immediately commences boiling, not so violently perhaps on the instant, but sufficiently so to change it from its hitherto quiescent condition to a state of active circulation at least, if not of actual ebullition at the surface. Now this or *any other disturbance of the surface* of the water, starting the engine for instance, will rapidly supply the surcharged steam with its full complement of water, that being all that is wanted to cause the pressure to mount up suddenly from 20 to 200 or 300 lbs., perhaps in a few seconds, or to such pressure as is due to the final temperature of the steam when fully saturated with water, the *dry* surcharged or *desiccated* steam, as it may be called, suddenly becoming ordinary steam of somewhat less temperature, but enormously increased in density and pressure, with what effect on the boiler of course depends entirely upon the strain it is capable of bearing without rupture.

Supposing the iron to have arrived at about 400° Fahr., fully saturated steam in contact with it will assume a pressure of about 215 lbs. per square inch above the atmosphere, a pressure quite equal to account for many of the disastrous explosions we have on record.

Without going farther into the subject in this place, if I

have not made it appear very probable that explosions sometimes do take place from the cause stated, I trust that the
possibility of their doing so is sufficiently evident, which is all
that was intended ; for there are many other causes concerned
in producing a state of things liable to lead to explosions
which require to be treated on after we have considered the
proper thickness and strength of iron to withstand any given
strain. My present object is mainly to induce a more con-
siderate attention being paid to the subject of properly hanging
the direct-draft boiler, referred to by Rule 1, so as entirely to
avoid any even remote possibility of danger arising from the
causes pointed out.

With this view, also, the following figures (10 and 11) are
given to illustrate the method I have long recommended and
practised of setting up boilers of this kind ; and although they
amount to some hundreds, I have never yet heard of one that
has exploded. I may also conclude with reiterating the
caution which it was my object to urge at the commencement
of this section, namely, that *the flame should never be allowed
to act against the sides of a cylindrical boiler above the level of
its centre,* much less above the level of the water.

Section 29.—Fire and Flame Bridges.

It is very important that all boilers of any considerable
length, and particularly when erected on the direct-draft plan,
should be provided with several flame bridges under them,
commonly called " *check* bridges," but perhaps improperly so,
from the supposition that their proper office was only to check
or impede the too rapid current of hot air and flame in their
passage to the chimney, and consequently to retain the heated
gases longer under the boiler, which they certainly do quite as
effectually as causing the smoke to travel through long, narrow,
tortuous flues. This, however, is the least important purpose
they subserve.

Fig. 10.

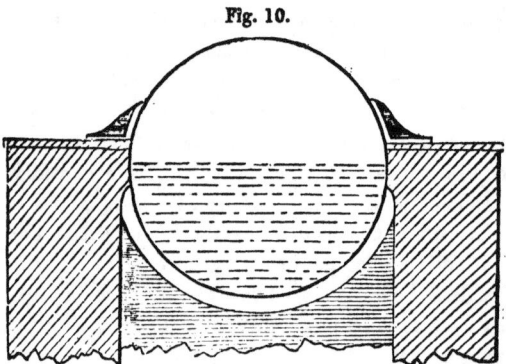

The above, fig. 10, is a transverse section of one of a number of boilers which were erected by a firm in Manchester several years ago, all of which are yet in work, showing how the boiler is hung upon cast-iron brackets, rivetted to the boiler a little above its centre, and resting by broad flanges on the top of the side walls.

This boiler was purposely chosen of this simple, and what may be called rudimentary plan, and put up in the cheapest and simplest manner with a direct draft, so that any alterations or improvements that it might have been found expedient to make, either in the setting or the construction of the boiler, might be in the shape of additions merely, and therefore capable of being separately proved, both as to first cost and actual worth ; and also that observations might be made upon it for a sufficient length of time, without the liability of error either from complication of construction or from interruptions owing to the necessity of stopping to clean out the flues or otherwise. It was thus made to answer the purpose of a trial boiler, in order to guide the firm to which it belonged in their choice of the kind of boilers to adopt in the erection of some new works.

One of the flame bridges is shown in elevation in the preceding figure (fig. 10). It is an inverted arch, 5 inches from the boiler bottom, and equally distant all round. Too much attention cannot be paid to the proper construction of these bridges; for neglect in this matter has always been the cause of any great waste of fuel that has ensued on putting up a direct-draft boiler. If too great a space is left above them, it is almost as bad as if the bridges were left out altogether; for then the flame is apt to divide itself into two currents, one on each side of the boiler, and thus run off to the chimney without taking much effect upon the boiler bottom.

I believe I was the first to design boilers with several flame bridges of this description, at least to publish an account of them, and urge their general adoption as an absolutely essential requisite to all boilers set up on the direct-draft plan, which I did on the ground that their proper office is principally for the purpose of spreading the flame and heated air around the convex heating surface, so as to completely envelope the lower half of the boilers in a stratum of flame of comparatively equal thickness or uniformity.

Similar observations may be made with regard to the fire bridge represented in fig. 11. It is too frequently built by

Fig. 11.

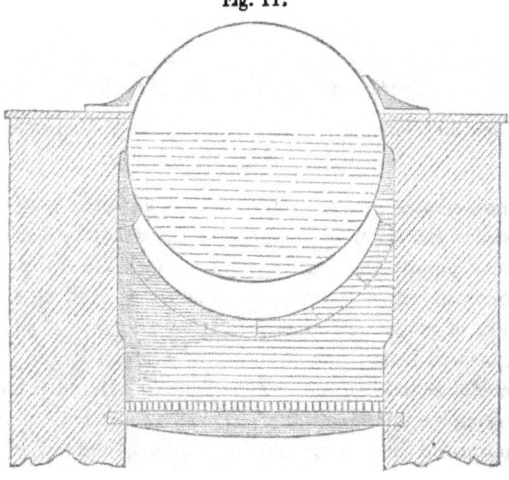

bricklayers in the form of a horizontal wall of very little elevation at the end of the fire grate, in evident ignorance of what the proper functions of a boiler bridge consist, and as is also evidenced by the name of "*stop*," or "fire stop," that is commonly given to this bridge, from the supposition that its only use could be to prevent the stoker from pushing the coals over the end of the fire grate. Its most important object, however, is, like that of the flame bridge, to act as a dam for the current of flame and gas to flow over.

The top of this fire bridge is described by a circle of the same radius as the boiler, at about 10 inches below the boiler bottom, as shown in the above drawing, which is a vertical section of the furnace across the back end of the fire grate on a scale of one-fourth of an inch to a foot. This fire bridge reclines backwards, with a batter of about 6 inches in the middle, diminishing to each side of the furnace, where the upper part of each wing of the bridge is vertical.

CHAPTER V.

On the proper Thickness and Strength of Boilers for Durability and Safety, with Examples of actual Cases.

SECTION 30.—PRACTICAL LIMITS TO THE THICKNESS OF RIVETTED BOILER PLATES.

THE proper strength of boilers, to enable them to withstand with safety the required pressure of the steam is a matter of such very great importance as regards both life and property, and the responsibility of the proprietors as well as the constructors of boilers are consequently of so grave a character, as might well justify the devotion of a much larger space to this subject, even in a rudimentary treatise. Happily, however, the principles on which the strength of all boilers depend are comprised in a very narrow compass, and, of whatever material the boiler is made, may be stated in few words, the strength being directly as the thickness of the metal and inversely as the diameter of the boiler.

So long as boilers continue to be made by hand labour, and the quality of iron remains what it is, the thickness of the wrought-iron plates of which they are constructed is also practically determined within exceedingly narrow limits. A good boiler cannot be so made *less* than a quarter of an inch, nor much *more* than half an inch in thickness.

Some of the best operative Staffordshire boiler makers are known to be even disposed to narrow these practical limits still more, or to $\frac{5}{16}$ and $\frac{7}{16}$, and do not hesitate to declare that with the average of workmen the proper thickness of boiler plates must be confined between a quarter of an inch, which cannot be *properly caulked*, and half an inch, which cannot be *properly rivetted*.

Without wishing to contravene this opinion, so far as the inferior limit is placed, it is very probable that the other and more important limit may be extended to half an inch or $\frac{5}{8}$ with the partial assistance of the rivetting machine, combined with properly paid and therefore superior hand labour. It is admitted that $\frac{3}{4}$-inch plates can now be better rivetted by the rivetting machine than $\frac{1}{2}$-inch can be done by hand. Useful, however, as is the rivetting machine, and important a part as that invention is yet destined to play in the history of boiler making, ship building, and *bridge* building, even little and apparently insignificant a step as it is beyond the boiler maker's punching engine, it has already done wonders; but it is yet only calculated for plain work, or large tubes. We have not yet got a machine that will go through the manhole and close up the end of a boiler.

Admitting the practical data contained in the foregoing observations, which I am compelled to do after a recent examination of numerous boilers in Staffordshire and the other iron counties, where it is common enough to find many of the reprehensible practices mentioned in Section 11 (p. 32) still

F

prevailing—we have scarcely any choice left but to fix on either $\frac{3}{8}$ or $\frac{7}{16}$ inch plates, as the thickness at which we shall most probably continue to get the best boilers made by hand. Taking everything into consideration, in answer to the frequent question of the best thickness, I usually say without hesitation $\frac{7}{16}$ for Staffordshire and $\frac{3}{8}$ for the best Yorkshire plates, may be fixed on, as it is generally allowed there is that proportionate difference in the strength of the two irons; and with the assistance of the rivetting machine the maximum thickness may soon be extended to $\frac{9}{16}$ of an inch.

In recommending $\frac{3}{8}$ iron generally, say for high-pressure boilers of small and low-pressure boilers of large diameter, as the least thickness that ought to be permitted for the shell of any kind of boiler whatever, there are other reasons besides its being the ordinary thickness for securing the best workmanship,—the principal one, in my estimation, being that, owing to its being so much used, we know well what it will bear. We know for instance that a locomotive boiler of $\frac{5}{16}$ thick and over 3 feet in diameter, will bear a working pressure of 80 to 100lbs. per square inch. I have also seen a boiler of 10 feet diameter and $\frac{3}{8}$ thick, tested with a cold-water pressure of nearly 100lbs. per square inch, which gave no indication of weakness, although I would very much doubt the prudence of working such a boiler at 50lbs., and more so after its being tested to that extent than before.

It is very commonly stated that boilers should be tested to *three* times their intended working pressure; *double* pressure is, however, quite ample, at least for high pressure boilers, but there ought to be no leakages at that pressure. Some also recommend the propriety of testing new boilers, in the first instance at least, with air instead of water pressure; and, as it is quite as important to have the seams as well caulked inside as out, the use of air instead of water would give some practical facilities for that purpose.

SECTION 31.—RULES FOR PROPORTIONING THE STRENGH TO THE PRESSURE AND STRESS ON THE IRON.

The tensile strain that good wrought iron is capable of undergoing without rupture, is so perfectly enormous that few people are able to believe that steam boilers are ever actually burst with the *fair* pressure of the steam, however great that may be ; operative boiler makers in particular are all strongly impressed with this idea, hence they are all staunch advocates of the various theories that it is *gas electricity*, or anything but simple steam, that is concerned in tearing a boiler to pieces.   Natural as it is, on witnessing the astounding effects produced by the uncontrolled power of steam, to ascribe them to occult, imaginary, and other far-fetched although so-called scientific causes, it is very blameable when parties, who ought to know better, encourage such belief, because it indisposes men who have the care of boilers from giving sufficient attention to simple palpable facts and circumstances in practice, by which, with the aid of a little common arithmetic, it is probable that all the accidents that have hitherto happened to steam machinery might easily be explained.

The usual rule for estimating the pressure that may be safely put on the cylindrical part of a boiler, is to multiply the number of pounds per square inch section that you will allow the iron to be strained to by the thickness of the plate in inches or fractions of an inch, and this product divided by the internal diameter of the boiler in inches, will give the number of pounds per square inch pressure that each side of the boiler must bear, in order to subject the metal to the given strain. Or, in fewer words, the separating force of a cylindrical boiler is equal to double the strength of the iron, the strain being borne equally on each side.

If we fix on 5000 lbs. per square inch section as the greatest allowable strain that iron should be exposed to, which is about one-third of the maximum strain at which it is liable to suffer a permanent derangement of structure, and less than

$\frac{1}{10}$ of the ultimate strength of the best wrought iron—the above rule may be expressed as follows :—

RULE 1.—Multiply double the strain allowable on the iron or (2 × 5000 =) 10,000, by the thickness of the boiler plate, divide the product by the diameter of the boiler, all in inches, and the quotient is the pressure of steam in pounds per square inch that the boiler will bear without injury.

Let s = the strain on the iron in lbs. per square inch section.
  t = the thickness of the plate.
  d = the diameter of the boiler.
  p = the pressure at which it may be worked.
Then the equation for the pressure is—

$$p = \frac{2\,st}{d};$$

and the general formula for our short slide rule is as follows :—

A	$2s$ = 10,000	d = diameter in inches.
Ɔ	t = thickness	p = pressure in lbs.

Example 1.—Required the pressure of steam at which to work a locomotive boiler of 36 inches diameter and $\frac{5}{16}$ inch thick, so that the greatest strain per square inch, sectional area, of the iron shall not exceed 5000 lbs.

A	$2s$ = 10,000	d = 36 inches.
Ɔ	$t = \frac{5}{16}$ = ·3125	p = 86·8 lbs.

Example 2.—A cylindrical boiler of common Staffordshire iron is 6 feet in diameter, and the circular part of the shell $\frac{5}{16}$ inch thick. It is required to find the pressure of steam it may be worked at, so that the maximum strain on the iron per square inch, sectional area, shall not exceed 3000 lbs.

In this case the gauge point for the strength of the iron is 2 × 3000 = 6000, instead of 10,000, and the operation is as follows :—

△	$2s$ = 6000	d = 6 × 12 = 72 inches.
Ɔ	t = ·3125	p = 25 lbs. per square inch.

Example 3.—The above is an actual example in practice of a badly-proportioned high-pressure boiler. It is required to find the diameter of a good one, capable of working at double the above pressure, or 50 lbs. per square inch, made of the best Staffordshire plates of the same thickness ($\frac{5}{16}$ths), but able to withstand a strain of 4000 lbs. per square inch, sectional area, of metal.

A	$t = \cdot3125$	$p = 50$ lbs. pressure.
ɔ	$2s = 8000$	$d = 50$ inches diameter.

Example 4.—A cylindrical low-pressure boiler is 9 feet diameter and 20 feet long, made of the best Yorkshire iron $\frac{3}{8}$ths thick : what is the pressure it may be worked at in order not to subject the iron to a greater strain than 5000 lbs. per square inch sectional area of the metal; and what is the nominal horse power of the boiler with Lancashire coal, the boiler containing two inside flue tubes on Mr. Fairbairn's patent construction, each of 3·33 feet in diameter?

First, for the required pressure :—

A	$2s = 10,000$	$d = 9 \times 12 = 108$ inches.
ɔ	$t = \frac{3}{8} = \cdot375$	$p = 34\cdot72$ lbs. pressure.

Second operation for the horse power, according to the formula, Section 5 (page 9), using 5 for a divisor or gauge point, which gives Boulton and Watt's proportions, or about 8 square feet of effective heating surface per horse power :—

A	$d = 9 + 3\cdot3 + 3\cdot3 = 15\cdot6$ feet	G P $= 5.$
ɔ	Length $= 20$	H P $= 62\cdot6.$

When a boiler is to be made of the very best iron, requiring the coefficient for its maximum strain, $2\,s$, to be fixed at 10,000 as above, it suggests a very simple arithmetical rule for finding the thickness for any given pressure, as follows :—

RULE 2.—Multiply the diameter of the boiler, in inches, by the given working pressure in lbs. per square inch ; then the 4 right-hand figures of the product will represent the proper thickness of the boiler plate in decimals of an inch.

In the last example we have—

Diameter of boiler	. . .	108 inches.
Pressure per square inch	.	35 lbs.

$$540$$
$$324$$

The required thickness = 0·3780 inches.

The converse of this operation is also useful ; that is, when the thickness of the plate is given in decimals of an inch, add to it 4 ciphers, and divide by the pressure for the diameter, or by the diameter for the pressure, as follows :—

The given thickness, $\frac{3}{8}$ = ·375 inch.

Diameter, 108 $\begin{cases} 9)3750 \cdot 000 \text{ with 4 ciphers added.} \\ \overline{} \\ 12)416 \cdot 666 \end{cases}$

34·72 lbs. pressure.

SECTION 32.—FAIRBAIRN'S PATENT DOUBLE-FLUED AND DOUBLE-FURNACED BOILER.

The 9-feet cylindrical boiler, in example 4, is an actual case in practice of a 60-horse boiler, which I designed for driving one of Boulton and Watt's 60-horse engines at the Hope Mills in Manchester, belonging to Messrs. George Clarke and Co., in the year 1844, and which was executed by Mr. Fairbairn, and put to work the following year with very great success, together with another boiler of the same kind of still greater proportions, namely, 10 feet diameter and 24 feet long, which has been already alluded to (Section 30, p. 96) as having been tested to nearly 100 lb. per square inch pressure. Either of these boilers are able to work the engine up to its full nominal power with very easy hand firing, and, when both boilers are working together, still more economically,

which they have generally done regularly, the engine being
loaded up to about 145 gross indicated horse power, and using
not more than 6 lbs. of very inferior coal per indicated horse
power per hour.

Having had the advising and designing the proportions of
these boilers, I need not say they would have been preferred
longer with the same diameter, but it was compulsory to
place them in a very circumscribed space of ground in a corner
of the factory yard, and much against the opinion of Mr.
Fairbairn himself, who had just patented these boilers and
who did not then approve of a greater diameter than 6 or 7
feet, and recommended a much greater length. As very many
of these double-flued boilers are now in use in Lancashire as
well as other parts of the kingdom, I make no apology for

Fig. 12.

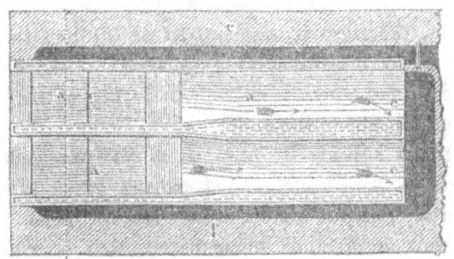

giving the drawings (figs. 12 and 13), showing a general plan

Fig. 13.

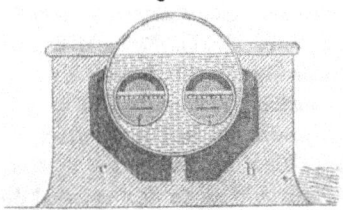

or horizontal section of the boiler and flues, as well as a transverse section of the same.

The particulars of the 9-feet boiler are as follow:—The cylindrical part of the shell, as well as the flat ends of the boiler are made of $\frac{3}{8}$-inch iron; it is 9 feet in diameter by 20 feet long, and contains two flue tubes, each 3 feet 4 inches diameter, but rather deeper at the front ends, which contain the two furnaces. The flues pass through the water space the whole length of the boiler, and are made of plates $\frac{7}{16}$ thick, or $\frac{1}{16}$ thicker than the shell, except the tops of the furnaces, which are of Low moor iron, and only $\frac{3}{8}$ thick. There is suffi- cient space between the flues, which are 13 inches apart, for a man to get through. The flat ends are braced together as usual, and also have four additional oblique stays at each end, radiating to different points of the upper half of the boiler; these, together with the two flues in the lower half, pretty well equalise the internal strain on the different parts of the boiler. It was intended to work at about 15 or 16 lbs. per square inch, therefore it was proved at 30 lbs., that being about $\frac{1}{3}$rd of its maximum strength, which it stood without any deflexion, and which is only about 1-10th of the ultimate or bursting strength of the boiler.

Although, as appears by example 4, this was called a 62- horse boiler, it only contained 54 square yards of effective heating surface; and it was found to work the most economi- cally at about 50 nominal horse power, that is, evaporating about 50 cubic feet of water per hour; the evaporation then being at the rate of about 8 lbs. water to one of coal, the coal being one half "Burgey" at about 6s. per ton.

The above described boilers were the first that were made of that large size, after Mr. Fairbairn took out a patent for them; and, as their introduction formed a very important era in the history of working factory engines to a greater extent expansively than previously had been usual, I was at some pains to investigate all the facts relating to them. It may be observed that it was only as to the external dimensions, the heating surface, and capacity to ensure their efficiency and economy in fuel, that I am accountable for, in these boilers;

the thickness of the plates, mode of staying, and consequent strength, being entirely left to the responsibility of the maker, who is so well known as an authority in such matters; my only care in this respect being that, as they were intended to work under 20 lbs., that pressure should not be exceeded, by causing the feed water to be supplied by the ordinary open stand pipe of about 50 feet high, so that the population of the factory, under a portion of which the boilers were placed, should not be dependent on the safety of the "safety valve" alone.

A more particular description of this kind of boiler is given by Mr. Fairbairn himself, in the British Association Reports of 1844, in which its smoke-consuming qualities are spoken of, respecting which it need only be stated, that in this particular case, as in all other cases, with a good fireman, and a good draft capable of burning up the whole of the coal with little or no stoking, there is little or no smoke. But although, from the shortness of the boiler, in proportion to its diameter, so completely opposed to the Cornish system, it was predicted as usual that " *all the heat would go up the chimney,*" its general economy was all that could be wished. And the earliest opportunity was taken of comparing its economy with that of another boiler of the same kind, of the same diameter and same size of flues, but of 28 instead of 20 feet long, for which I was requested to make designs for setting up, at another cotton factory in the same neighbourhood. This boiler was made by Messrs. John Petrie and Co., of Rochdale, the same year, and set up to work one of their 40-horse patent expansion engines, at Messrs. Hugh, Shaw and Co's.; and, when using coal of a similar quality to the former, was found to have as nearly as could be ascertained the same proportion of evaporative economy, the economy of working with a variable amount of expansion at this particular engine being the subject of a series of careful experiments, which were published at the time, April, 1846. The facts, however, in relation to the economy of the boiler, which is all that now concerns us, were sufficient to show that nothing is to be obtained by making a boiler of this kind longer than about 3 times its diameter, which agrees with the general rules on the subject laid down in the last chapter. (Sect. 27.)

Notwithstanding the only conclusive proof in all matters physical, namely, experiment, is without exception in favour of the general doctrine I have endeavoured to inculcate in respect to the length of boilers, and therefore all mere opinion may be safely repudiated, I will venture to state one practical man's authority on this subject, whose opinion that the current notions of danger of explosion by having boilers of large diameter, or of loss of fuel by want of great length, are mere prejudices, will not be disputed, when I mention the name of my friend, Mr. William Elsworth, C.E., of Preston; that gentleman having erected some boilers of this construction the same year, at the splendid cotton works of Messrs. Horrockses, Miller and Co., of 24 feet 6 inches in length, and not less than 10 feet 10 inches diameter, he agreeing with me that danger of explosion is more to be dreaded from bad materials, bad workmanship, and *small* boilers, which a man cannot get properly into, either to make or to clean, and keep in repair, than all other causes put together.

Boilers of 10 feet to 12 feet in diameter, made of the best Low moor iron $\frac{3}{8}$ thick, with the ends $\frac{1}{2}$ inch, and braced with three or four longitudinal stays of 2 inches square, may be quite safely worked up to 30 lbs. per square inch pressure, with a certainty that no part of the shell of the boiler at least is exposed to so much as $\frac{1}{3}$rd of the strain the iron is capable of bearing without injury.

SECTION 33.—STRENGTH AND FORM OF INTERNAL FLUE TUBES.

The weakest part of a boiler constructed on Mr. Fairbairn's principle, is undoubtedly the internal furnace or flue tube, which is notoriously liable to suffer from collapse, if made of thinner iron than the shell. It is undoubtedly true that the same mathematical law of the stress applies, as shown by both Emerson and Tredgold, to the external pressure on the flue tube as to the internal pressure against the shell, or that if $\frac{1}{4}$ inch thick is sufficient for a 6-foot shell, $\frac{1}{4}$ inch is enough for a 3-foot flue to sustain the same pressure, *so long as the latter retains its true circular figure.* Unfortunately, however, this is an *impossible* condition in a wrought-iron boiler, for the flue cannot be said to *retain* a figure which it cannot be made with at first, therefore it has been endeavoured to be met, as such difficulties too commonly are, by practical men, with a compromise between safety and danger—that is, making the thickness of the flue plates intermediate between that which is theoretically correct and that which their habits only have taught them to be practically wrong ; or in the above assumed case between $\frac{1}{4}$ and $\frac{1}{2}$ inch, namely, $\frac{3}{8}$ths. We thus find this to be a kind of universal thickness for nearly all flue tubes, whatever may be the pressure they have to sustain.

If it were not the fact that extensive loss of life is continually taking place from explosions, by far the greater number of which are known to commence, at least, from a collapse or giving way of the flue tube, this question might remain until the experience of each particular boiler maker settled the point to suit his own convenience ; but as it is, it is pre-eminently one of those questions that ought to be taken out of the hands of mere practical men so called, who are too often themselves but a compromise between good workmen and bad mathematicians, and settled by *law,* which ought to enforce the necessity of having all such structures " stronger than strong enough." Taking this view of the case, I have never seen any valid objections to making the inside flue of even thicker iron than the outside shell ;—at any rate, when a flue exceeds 3 feet in diameter the plates ought not to be less than $\frac{7}{16}$ inch thick, however low the pressure of the steam may be.

In flue tubes greater than 3 feet in diameter for high-pressure boilers, of course the proper thickness soon approaches the limits of good workmanship, and when that is reached, any system of bracing for supporting a flue should be resorted to with very great caution. Although we have many highly ingenious examples of bracing and staying in the low-pressure boilers of some of the American steam boats, after carefully examining several of them I have come to the conclusion that they ought not to be depended on to the exclusion of other considerations even for low pressure, and much less for high. Indeed, if very high pressure is ever to be justified in steam navigation at all, it would be better to return to Oliver Evans' and Woolf's systems of small water-tube boilers, than any further complication of Stephenson's locomotive tubular-flued boiler, which the present marine "tubular boiler" is in fact.

The safe principle of Oliver Evans' boilers, which is that of never allowing the pressure to be exerted except *within* cylinders of comparatively small diameter, appears to have been carried out in a very scientific manner by Dr. Alban, a practical steam-engine maker at Plau in Saxony. He has published the results of his experience in a work called "The High-pressure Steam-Engine Investigated,"* which are exceedingly important and interesting to all those who would concern themselves with steam of 200 or 300 lbs. pressure. And to those who have serious intentions of promoting aerial navigation *by steam*, I would say this is the direction in which you must look. It is perhaps not too much to say that Dr. Alban has constructed engines of great simplicity, and of greater power in the same space, and of less weight, consistent with durability, than any preceding engineer.

The best way of strengthening the large internal furnace flues of boilers is by rivetting on them a series of ribs of angle or Tee-iron at short distances apart, similar to the iron ribs used for the tops of locomotive fire boxes. This plan is the invention I believe of Mr. Joshua Milne, of Shaw, near Oldham, who has had greater experience than any other manu-

* A translation of this work into English is published by Mr. Weale.

facturer, in adapting the high-pressure expansion engine to cotton spinning. Safety to the hands employed being the chief requisite in using high-pressure steam for factory purposes, Mr. Milne's plan cannot be too widely known, for it may be easily carried out to such an extent as to make the collapse of a flue almost inconceivable.

SECTION 34.—MARINE BOILERS.

That part of the flue tube beyond the furnace admits of another mode of strengthening which is likely to be preferred by the very large class who, while studying safety, do not forget economy. The principle of the plan I refer to is that of using vertical *prop stays* in the flue in the form of tubes, through which the water of the boiler circulates, and against which the flame acts in its passage through the flue.

The above principle has been acted on for many years to a greater or less extent in this country, but recently more generally by the Americans in some of their large Atlantic steam boats. Those boilers on this principle which have been tried for steam navigation in this country have been very efficacious in generating steam, but remarkably addicted to *priming*, as in fact good boilers generally are, and on that account never came much into use. They contain, however, many of the elements of a good boiler for marine purposes, and only require the vertical water tubes to be *shortened* and *tapered*, so as to give free egress for the steam without undue or too rapid a circulation of the water, to prevent a good deal of the priming, even with the water of the muddy Humber, where they have been mostly in use; that river being, *par excellence*, adapted for priming, or "*fermenting*," as is the expressive term there applied. This is the universal malady of steam-boat boilers when leaving the clear water of the ocean and coming within the vicinity of *dirty fresh* water, which is at such times so conspicuously detrimental to the working of the engine and everything relating to it.

To say what is necessary, if we go into the subject of marine boilers at all, would very far exceed the present intended limits of this volume; their forms are so numerous and diversified, that to give an adequate description of only a few would be to do great injustice to many perhaps equally as deserving. Suffice it to say that the present position of marine boilers is one of transition, and they may be said to be in their *second* transition state. The old large-flued marine boilers, it is well known, were nothing more than land boilers placed in difficult circumstances, that is, on board ship, surrounded with water spaces instead of brickwork. That was their first stage; but marine engine makers still adhered most inveterately to the then prevailing prejudice of the scientific engineers of the day, the Cornish doctrine of long lumbering interminable flues—until some of them had the courage to break through the established routine by adopting the locomotive tubular-flue boiler of Stephenson, since called the *tubular* boiler, which designation properly belongs to the small *water-tube* boilers only. This second stage of improvement was effected by turning the tubes over the fire box or furnaces; hence they are called "turn over" boilers, which very great improvement I believe Mr. Seaward of London has the merit of effecting. Although this adaptation of the locomotive construction enabled the engines to be worked at higher pressure and greater expansion with a smaller boiler, there are many inconveniences in respect to their want of durability, from various causes which railway boilers are not exposed to, which render it very desirable that further *original* improvements should be made before much greater speed can be expected in steam boats. It is not difficult to predict that any farther mere *imitations* of the locomotive construction, respecting which our marine engineers are all at sea again, will be ineffectual in preventing our American brethren going ahead of us, and which some of their late achievements in this line give great indications of their doing. At any rate if the Americans have not already arrived at the best form of marine boiler, it is yet to be invented.

SECTION 35.—GALLOWAY'S PATENT DOUBLE-FURNACED TUBULAR BOILER.

Whatever turn marine engineering may take in the next few years, safety from explosions must always be a paramount consideration in these days of cheap travelling, which is my main reason for mentioning the subject of the last section. I shall therefore take this opportunity of referring marine boiler makers to a combination of the tubular and flue construction, recently brought a good deal into use as a land boiler, by Messrs. W. and J. Galloway, of Manchester, the principle of which may be easily adopted in steam boats.

It is not only the strongest form of boiler for its dimensions that has yet been made, but it is believed to be the most economical for its weight. as a generator of steam.

The following figures, 14 and 15, represent a longitudinal

Fig. 14.

and a transverse section of this boiler, showing the mode in which the main fire flue is supported and strengthened by a series of short vertical water tubes, which are made slightly conical, or about 2 inches wider at the top than the bottom, and amongst which the flame is allowed to play in its passage through the flue, the tubes being placed zigzag fashion, as seen in the horizontal section, fig. 16, giving great facility for this purpose. This particular arrangement of short water tubes, to cause them to act as prop stays of the strongest possible form, and in the best position for resisting any collapse of the fire flue, is the most valuable feature in Messrs. Galloway's invention as regards safety. In respect to the absorption of heat from the flame, the disposition of these tubes is also remarkably favourable ; for,—avoiding the difficulty of causing the flame to make its way through a crowded box full of comparatively small tubes on the one hand,—the tendency of the flame to divide itself into two currents, which a single row of

Fig. 15.

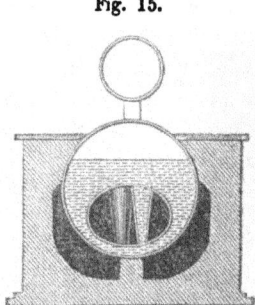

tubes sheltering each other would promote, is also prevented on the other. This arrangement also assists in causing the

Fig. 16.

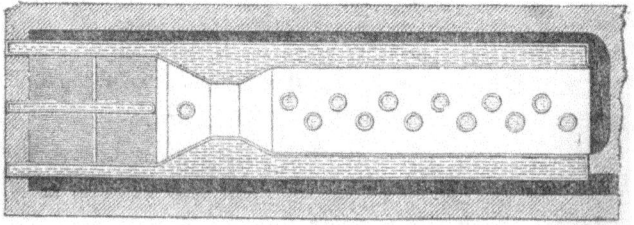

flame to wrap and envelope the tubes, so as to render a greater proportion of their surface effective. Hence this kind of boiler is pre-eminently distinguished for its great economy with flaming fuel.

Apart from the subject of the strength of boilers, we cannot help remarking on the utter disregard engineers generally have paid to the distinctive characters required in boilers adapted to flaming or non-flaming fuels. The locomotive boiler, for instance, is especially adapted for coke or other fuel, with little or no flame, nor was it ever meant to be otherwise. Its great author, one of the inventors of the safety lamp, could not be ignorant of the fact that flame would not enter far into narrow tubes; neither was he, for in one of the last interviews I had with him, which was on the subject of the present work, he fully agreed with me on the propriety of considering the locomotive boiler, so far as the small tubes are concerned, to be merely an apparatus for *heating with hot air*, and not at all adapted for the use of Newcastle coal. The only object to be obtained in passing the products of combustion through an immense number of very small tubes, is to drain out, as it were, the last dregs of the *caloric*, as Mr. Stephenson expressed it, after the great bulk of the steam is obtained from the action of the flame in the fire box, on which his chief reliance was placed. Yet, in defiance of this important practical distinction, we have seen English engineers pushing the small *tubular-flue* principle in marine boilers to the uttermost, for burning *bituminous* coal; while, on the other hand, we have heard of American and other engineers using the anthracitic and other non-bituminous varieties of fuel to the simple boilers of Oliver Evans and Trevithick, so much better suited to bituminous fuel or to wood.

Besides the arrangement and position of the water tubes in this boiler being well calculated for strengthening the flue and for intercepting the flame, there is another advantage not less important, as it equally affects both their durability and their efficiency in taking up or absorbing the heat from the passing flame, and that is the *conical form of the tubes.* I have already referred to the difficulty always experienced in causing the heat to pass laterally through the vertical sides of boilers with sufficient rapidity for generating steam, and which in fact is so great as to justify us in considering the effectiveness of a perfectly vertical surface to be only about one-half that of an equal area of surface inclined even at a very small angle towards the horizontal; the reason being the intervention of the rising bubbles of steam, preventing the necessary contact of the water at the upper portion of the surface; such vertical position of the heating surface also causing the metal to become over-heated, just in the same proportion as it is prevented communicating that heat to the water. This difficulty in the lateral communication of heat is obviated by the enlarged surface of the upper ends of the tubes in a certain measure inclining over the flame. The current of flame is also, by the same means, confined principally to the lower part of the flue without the intervention of *hanging* bridges, which are always necessary with perfectly vertical surfaces or tubes of a uniform diameter.

I formerly took occasion to call the attention of boiler makers to the evident advantage to be derived from causing the heating surface thus to *overhang* the source of heat, more particularly as respects the sides of the furnaces of marine boilers and fire-box boilers generally. And I have since had the satisfaction of finding my advice acted on with considerable success, both in land and marine boilers. Although this boiler of Mr. Galloway's is the first in which the principle has been applied to vertical *tubes*, it has given some opportunities of proving in practice that the application is still more efficacious in tubular than had been previously found in flat surface.

Section 36.—Boilers at the Gutta-Percha Works.

As a recent application of those boilers in London, and on the principle that example is better than precept, I have the kind permission of the proprietors of the Gutta Percha Works, in the City Road, for stating that on their consulting me on the propriety of repairing or reconstructing several of their boilers, I did not hesitate to take the responsibility of strongly recommending the erection of a new 50-horse boiler, on Messrs. Galloway's plan, which recommendation was adopted, and with such success in respect to economy in fuel as to induce the Company since to erect another boiler of the same construction and dimensions. This last boiler is represented by the above figures, 14, 15, and 16 (pp. 111, 112). To the description there given it is necessary to add that the great reputation these boilers have already acquired in Manchester and other places, on account of their being good " smoke burners," and where very strict enactments on the subject have made it compulsory, is one reason for their now being adopted by some of the large brewers and others in London; and the plan by which this desirable object is accomplished is most decidedly the simplest, and is on that account perhaps the best, that has yet been tried. It is simply by firing each of the two furnaces alternately, and allowing a certain time to interlapse between each firing proportioned to the quantity of coal laid on. Thus the fire from one furnace consumes the smoke of the other without the necessity of admitting air through any other than the ordinary openings between the bars, and without any extra trouble or attention of the fireman, he not having to attend to the opening and shutting of valves, nor to any machinery whatever.

Treating now of the strength of boilers, I may state that great strength is attained by this plan of double furnaces, while the effective surface exposed to the radiant heat of the fire is by the same means increased. The furnace grates are each 7 feet 4 inches long, by 2 feet 6 inches wide, making 37 square feet area, or at the rate of $\frac{3}{4}$ square feet of grate per horse for the 50-horse boiler. They each contain two sets of fire bars, the front set being 1 inch, and the back set 1$\frac{1}{4}$ inch thick, the draft spaces being $\frac{3}{8}$ inch wide in both. The furnace tubes are oval, being 2 feet 9 inches deep, by 2 feet 6 inches wide, but they are stronger than if they were circular, on account of the grate bearers acting as prop stays. The furnace plates are of Low moor iron $\frac{3}{8}$ thick. The flue and shell are of the same thickness, and the ends $\frac{1}{2}$ inch. The working pressure is from 30 to 40 lbs. per square inch, the mean of

which gives a strain of 4000 lbs. per square inch on the
sectional area of the iron.   The ends of the boiler are
strengthened by wrought-iron double ribs, which are rivetted
upon them, reaching from side to side, and to which the
wrought-iron stays are attached that extend the whole length
of the boiler and brace the two ends to each other.   The total
length of the boiler is 30 feet 3 inches.   The greatest diameter
of the main flue is about 4½ feet, and it contains thirteen of
the conical water tubes, each being 11 inches inside diameter
at the top, and 9 at the bottom.   The total *effective* heating
surface of the boiler, measured by the rules already given, is
rather more than a square yard per horse power, calling it
50-horse.   The makers call the boiler 55-horse, which, with
an adequate area of fire grate, it is quite equal to.   In order,
however, to make it a perfectly smokeless boiler with Newcastle coal, it is better to work it within its maximum power,
the rate of combustion being not more than 10 lbs. per square
foot of grate per hour.   At this rate it is capable of driving
an ordinary 30-horse non-condensing engine, loaded with
machinery to indicate on the average 50-horse power gross,
that is, including the friction of the engine itself, besides supplying at least 10-horse power of steam for other purposes.
With this load on the engine the evaporation is very accurately
1 cubic foot per minute, and the consumption of coal, not of
the best quality, called East Adairs Main, is at the rate of
3 cwts., or 336 lbs. per hour; this is the ordinary *net* consumption, that is, exclusive of what is required for first getting
up steam.   This gives an evaporative economy of between 11
and 12 lbs. of water for each lb. of coal consumed; a degree of
economy, I believe, that has been rarely, if ever, before obtained.   It ought to be stated that the boiler is well clothed
with a non-conducting covering of sawdust and brickwork,
and surmounted by a horizontal steam dome (also felted),
12½ feet long by 3½ in diameter, so that no error can have
place in the above data from priming, and the quantity of
water evaporated was ascertained by stopping off the feed and

measuring the depth which the surface of the water fell per hour in the glass-tube water gauge on the front of the boiler.

CHAPTER VI.

On Explosions, Deposit of Sediment, and Incrustations.

SECTION 37.—EXPLOSIONS FROM COLLAPSE OF FLUE.

THERE are only two ways in which a boiler can be caused to burst or explode by the power of steam. One is by a gradual increase of pressure produced in the usual way, but at a time when all egress is prevented, until the steam acquires sufficient strength to rupture the material of which the boiler is made ; the other is by some *sudden* increase in the quantity or pressure to such an extent, or with such rapidity, that the ordinary safety valves, or perhaps any other means of outlet that might be devised for the purpose, are unable to carry the steam off in time to prevent any, although but momentary, strain greater than the boiler can bear. I have long been of opinion that it is in the consideration of the last-mentioned class of causes that we ought to look for the proper remedy against the frequency of steam-boiler explosions.

The method of giving to boilers sufficient strength of material to resist any given pressure within certain limits was the subject of the last chapter, and disposes of that part of the question relating to the strength of the boiler while it retains its proper form ; so that I shall now only refer to the facts of some explosions in order to elucidate the causes of change of form taking place.

There is a very erroneous, although prevailing opinion, that the Cornish or Trevithick boiler, by having the fire place within its internal tube, is safer than any other kind, which opinion cannot be too soon dispelled; for it is an admitted fact, by all who have considered the subject, and however they may differ as to the precise theory of its action, that the water accidentally getting too low is a frequent cause of explosion;

and if so, it must be evident that this cause must operate much more frequently to produce that effect when, as in the Cornish boiler, the depth of the water over the hottest part of the heating surface is only a few inches, than when the depth is as many feet.

The force of the steam and water escaping during an explosion of a Cornish boiler is also immensely increased by reason of its being generally expended in one direction, that is, through the fire place in the mouth of the tube, the latter being thus converted into an immense piece of ordnance, from which the grate bars, fire bricks, and other materials of the furnace, are projected with destructive effect on everything within their range. Probably also the steam, as it rushes out, is reinforced by contact with the heated fuel in the furnace.*

Boiler explosions of this kind generally arise from a collapse of the internal flue, and it frequently happens that some of the most violent and fatal ones have occurred without the circular part of the shell of the boiler itself being in the least disturbed or removed from its place. Such were the two fatal explosions that occurred in the Thames on board the Victoria, Hull steam ship, in March and June, 1838, the long investigations respecting which are well known.

An explosion of a nearly similar kind to those in the Victoria steamer, both as to cause and fatal consequences, also took place in September of the same year, with a Cornish boiler, in a foundry at Newton in Lancashire, although, from its belonging to a high-pressure engine, the force of the explosion was much greater. Several tons in weight of cast iron and other articles were removed by it, and a clear breach was made by them of ten yards wide through the yard of the foundry, whilst everything in the direction of the mouth of the flue tube, for sixty or seventy yards in a direct line, and several yards wide,

---

* There are other circumstances which have been observed in the bursting of a boiler of this kind, which bear some analogy to the discharge of a large cannon or mortar; such, for instance, is the sound or report produced, which is not so great with boilers that have no internal flue.

was swept away with terrific violence, including ten or eleven
of the workmen, nine of whom were killed. The bricks of
which the fire bridge had been formed within the tube, were
projected like shot from a gun to twice the above distance, and
were the principal cause of the loss of life.

The above instance, in which the report was described as
like a loud clap of thunder, is adduced as an illustration of the
peculiar violence incidental to an explosion of this particular
kind of boiler, owing to the steam being reinforced, as it were,
within the furnace tube, and then being all expended in one
direction.

I had occasion to witness the melancholy effects of this
accident, and took some pains to form a mature opinion as to
its cause on a careful investigation of the facts, which were
published at the time. I also soon after, namely in 1839,
published and strongly recommended the only means that has
yet been proposed for avoiding the faulty construction which
led to the disaster, in this instance, by a collapse of the flue
tube.

The peculiar fault of this boiler, and the proximate cause of
its bursting, was that the tube was *oval* in section, although
the boiler itself was circular, a fault which, for many reasons,
is likely to be more productive of explosions than formerly,
and is therefore deserving the particular consideration of the
users and makers of *small high-pressure* boilers.

Now the main object in making an inside flue oval, and
placing it with its shortest diameter vertical, is no doubt for
the purpose of obtaining a greater depth of water over the
flue without diminishing the heating surface or the amount of
steam room ; but by thus endeavouring to avoid the chance of
accident from a deficiency of water, we run into the contrary
extreme and risk an explosion by making the flue of a weak
form. A very slight departure from the true circular form
not only causes the flue to be much weaker, but the pressure
has a constant tendency still farther to alter the form of the
curve, thereby becoming weaker with every strain until a

collapse of the flue takes place. The top and bottom of the flue are thus sometimes crushed flat together, or nearly so; and the rupture producing explosion consequently usually takes place in the flue itself, through which the steam and hot water are discharged in the manner already stated.

Fig. 17.

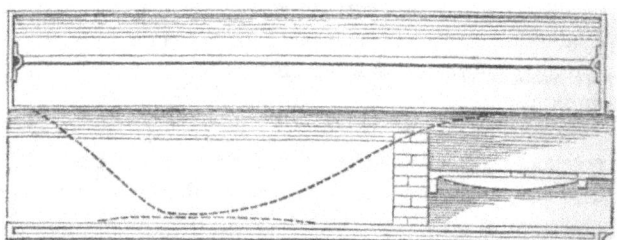

The boiler, of which the explosion has just been described, is represented in fig. 17. It was 12½ feet long by 4¾ diameter. The flue tube was 3 feet wide by 2½ deep. The brick fire bridge was at about one-third the length of the flue, and the top and bottom of the flue were crushed together at about midway between the farther end of the boiler and the bridge; the latter no doubt, interposing as a momentary support to the top of the flue at the instant of the plates coming down, determined the place of the collapse, as marked by the dotted lines, which show the exact form which the collapsed flue assumed by the force of the explosion.

This boiler was quite new, the explosion having taken place the first morning it was set to work, and within one minute after starting the engine. The plates were ⅜ inch thick and, saving the *form* of the flue tube, the boiler was remarkably well made as well as all the apparatus belonging to it. It had *two safety valves*, two gauge cocks, and a glass water gauge. The foreman of the works, who had the superintendence of erecting both the boiler and the engine—which was also new and of 8-horse power—was present, and managing them himself when the explosion took place, he being also one of the unfortunate sufferers.

The coroner's jury returned a verdict of accidental death occasioned by the insufficiency of water in the boiler, in concurrence with the opinions of several most respectable engineers, but with the addition of some of them ascribing the explosion to the *sudden formation of hydrogen gas*, by the injection of cold water upon the *supposed* red-hot flue, of which there was no evidence, when the engine started. Which last opinion is far from being a singular one in many similar cases of explosion that have occurred with high-pressure boilers ; but is I think a very erroneous one, not to say fatally so, in many instances. For by thus assuming a theory which, to say the most for it, is, according to our ordinary knowledge of the laws of chemistry, exceedingly improbable, a check is placed upon any further investigation, while the real errors of construction are perhaps kept out of view or repeated in other cases. The great fault in the construction in this case was evidently *too wide*, and therefore too weak, a flue in order to admit a large fire place in *too small* a boiler, which naturally led to the error of making the flue oval. The best preventive of this in the first instance would have been either a larger boiler or a separate steam chamber, as shown in figs. 14 and 15 (pp. 111 ,112), without resorting to any flattening of the flue.

Now, if instead of the flue the *boiler itself had been oval*, with its longest diameter vertical, and the *flue circular*, the same means of obtaining a greater depth of water over the flue would have been afforded, but with much greater safety from explosion by collapse of the flue. Although an elliptical or oval shape is slightly weaker for a boiler than a circle of a mean diameter, still, from the pressure being exerted *inside the curve*, any extra strain or pressure that the boiler may be exposed to will only have a tendency to alter the curve into a stronger form than it was before, by approaching more nearly to the circle. Thus, leaving out of consideration deterioration from wear and leakages, it may truly be said of an oval or egg-shaped boiler that *the longer it is worked the stronger it will become.*   On the other hand, when the pressure is *external to the curve,* as is the case with the flue tube, the effect is exactly the reverse of the above, the strain having a constant tendency to put the curved surface into a weaker position, and it must inevitably become weaker and weaker with every undue strain it is exposed to.

Consequent on the above considerations, it ought to be a rule in the making of high-pressure boilers, that the inside flues, besides being circular, should have their plates quite as thick, if not thicker, than the external shell.   The boilers themselves may be to a considerable extent elliptical, and will be, for all practicable purposes, as strong as if they were circular.   They have even been made as much as 8 feet diameter by 9 feet deep, without materially diminishing their ultimate strength; although for very high pressure, say above 30lbs. per square inch, I think the circular form should not be departed from when of so large a diameter.

Section 38.—Collapse of Flue in a Low-Pressure Fire-Box Boiler.

The better to illustrate all the causes and consequences of collapse of flue tubes, as well as explosions generally, I give the particular drawings in the frontispiece, which are correct representations of a boiler that I examined minutely immediately after its explosion, which occurred in Manchester a few years ago, being also well acquainted with all the circumstances connected with the boiler previous to its explosion; and I select this case because, although fatal to the fireman, and having many points of great resemblance to the *high-pressure* explosion at Newton, it is an example of as *easy* an explosion, so to speak, as can well be produced, even by *low-pressure* steam.   The steam in this case we are quite certain could not have exceeded 15 lbs. per square inch; in this respect, as in the immediate effect produced, being strongly contrasted with the Newton explosion.   This fact is one of some importance for the consideration of those engineers who are continually reasserting their belief in the dangerous fallacy that *high* pressure is as safe as *low*, and who do not hesitate to take the responsibility of recommending high-pressure engines in factories,—where in some very rare cases and circumstances they may be necessary,—as well as in steam vessels, where they never can be necessary or justifiable at all.

This boiler, which was on what is known as the fire-box construction, was, as respects materials and workmanship, remarkably well made and judiciously stayed, and will serve to illustrate that class of boilers.

The fire-box boiler is an offshoot from the locomotive, adapted to stationary purposes, as the Butterly boiler, which it closely resembles, is an offshoot from the Cornish, adapted for factory use,* the only variation in this instance being an adaptation of the furnace for consuming smoke by an arrangement which is not quite innocent perhaps of having a tendency to promote accidents of this kind,—a tendency which there is no denying that almost any plan of smoke consuming by a more intense heat or more perfect combustion must have *per se*. The particular arrangement for the purpose in this case, which has been frequently patented, is the interposition of a " hanging" or inverted water bridge at the back part of the grate room, descending within a few inches of the burning fuel, close to which the smoke from the newly laid on coal is compelled to pass, and thus become inflamed before passing into the mouth of the flue tube. Now it is not in this descending water bridge, provided it is kept clean, that any injury is generally done by this plan of smoke burning, as some suppose, but in the vertical plate at the back of the furnace, to which the mouth of the flue tube is rivetted, on account of the

---

* For a further genealogy of the fire-box boiler, and its variations under the name of the Liverpool patent construction, in 1832-3, afterwards successfully naturalized as a river steam-boat boiler in America, see my paper on the "Locomotive Engine Boiler," in vol. i. of Weale's new edition of Tredgold on the Steam Engine.

intense combustion created by the rapid draught of air that always rushes through the back of the grate. Consequently it was only this *vertical plate*, and *because it was vertical*, that showed any symptoms of injury from being over-heated, this overheating arising, as it must always do in all similar cases, from the difficulty the steam has in freeing itself from vertical surfaces, except by keeping the water out of contact with the plate.

The descending water bridge of this boiler is shown at c, fig. 2 of the frontispiece, which is a longitudinal section of the boiler, and in plan in fig. 3, which is a horizontal section; both figures showing by dotted lines the form of the flue tube in its collapsed state, but better seen in the transverse section, fig. 1, taken through the back of the furnace at B in fig. 2, which shows the fracture of the mouth of the tube where it was torn away from the angle iron that connected the tube to the front plate, and making the large aperture of 3 feet wide, through which the contents of the boiler rushed out. Fig. 4 is another transverse section of the boiler through A, showing a back elevation of the brickwork and flues as viewed from the rear.

The above figures are all to a scale of $\frac{1}{8}$ inch to a foot. The boiler was 23 feet long and $7\frac{1}{4}$ diameter outside; the flue tube was 16 feet long by 3 feet wide, and 30 feet deep at the back end, and of rather less depth, by 30 feet wide, forming a *very flat oval* at the mouth, where evidently the collapse of the flue first commenced, and where its connection with the back plate of the furnace might have been slightly weakened by the continued action of the flame, although the iron was not discoloured, neither was the *lead plug* melted, which was within a few inches of the place, as shown at the crown of the arch, fig. 1, and directly under B, fig. 2.

The thickness of the plates was $\frac{3}{8}$ inch throughout both the shell and the tube, and the vertical back plate was $\frac{7}{16}$ inch. All the flat parts were well supported by stays, as shown in the figures. The "water legs" of the boiler, as seen in cross section, were 12 inches wide, and supported by 6 "thimble stays" in each leg, of 1¼ inch round iron, with 3 of the same kind through the water space at the back of what *ought,* in all such cases *to be* a brick fire bridge. The ordinary working pressure was under 8 lbs. per square inch, at which the safety valve was adjusted to blow off, and the steam could not by any possibility reach to double that pressure before it boiled over at the feed head.

The total effective heating surface of the boiler and tube was about 44 square yards, and the area of the fire grate 34 square feet. It had a good draught, and in conjunction with another boiler of the same kind and dimensions, was quite equal to work a Boulton and Watt 53-horse engine up to 114 indicated horse power gross, with great economy. This they had continued to do for about two years after they were made, when the explosion took place. It occurred just after the engine started in the morning, and had taken about 20 turns. The effect of the explosion was to turn the boiler completely over longitudinally and laying it upon its back, without being entirely lifted from the ground. This movement was effected by the force of the issuing current of steam and water being deflected at right angles downwards by striking against the water bridge, thence driving out the fire bars against the bottom of the ash pit, and by reaction raising up the fire-box end of the boiler, which thus described a semicircle in the plane of the boiler's length, as already stated.

It may be useful to remark, in reference to the quiet way in which the boiler was turned over on its back, the contrast it affords to the high-pressure explosion at Newton; and in this respect it is still more strongly contrasted with two explosions of boilers of high-pressure engines, which also occurred near Manchester a few months afterwards, that is, early in the following year (1845). One of these was that of a locomotive at the Manchester and Leeds railway station, whilst getting up steam, by a collapse of the fire-box, the roof of which was crushed in with about 70 or 80 lbs. pressure, making a hole of about the same area as that in the low-pressure fire-box just described. The discharge of steam and water was also similar, that is, downwards through the fire grate; consequently the effect produced was likewise similar, but more violent, corresponding to the greater pressure of steam, the engine in this case being lifted entirely from the ground and making a kind of summerset in the air.

The other case was of still more violent character, being that of a Cornish boiler 26 feet long by 5¾ diameter, containing a fire tube of 3 feet diameter, the flue tube being collapsed for nearly its whole length from the furnace to the end of the boiler. The tube was at the same time broken across into four portions, which, when put together, exhibited very much the appearance of the collapsed flue of the fire-box boiler (fig. 2 of frontispiece), except that the top and bottom of the tube were compressed into closer contact and more completely recurved within each other. This probably arose from the greater pressure of the steam, the working pressure of the engine being 45 lbs., the thickness of the iron being the same as the low-pressure boiler, namely, ⅜ inch of the best Coalbrooke Dale, both for the boiler and tube. The tube in this case was *circular*, although it appeared to me to have been injured from overheating; and there were other circumstances which also made it probable that there was an insufficient quantity of water over the tube at the time of the accident, by which it might have been rendered quite as weak as the oval tube of the Newton boiler, to which case in many respects this bears great analogy, being both Cornish boilers and nearly of the same dimensions, except in length, and the explosion occurring within about one minute of the engine starting. In both cases the iron was of very superior quality, and the boilers extremely well made—provided with every requisite in the shape of safety valves, water gauges, &c., and attended by careful workmen. The principal difference in the manner of the explosion arising from the tube in the last case breaking across, the steam rushed into the internal flues and bed of the boiler, which served as an abutment from which the entire shell of the boiler, weighing several tons, was projected and carried a distance of 43 yards, knocking down buildings, and creating great destruction of life and property in its way. This, in fact, was the chief characteristic in the three high-pressure explosions described, as in *all other explosions of high-pressure boilers* that I have had an opportunity of examining,—which contrast remarkably with the almost entire absence of such destruction by the bursting of boilers ordinarily attached to the common low-pressure engine.

The reason for giving the details of the above cases is my belief that a correct record of facts is all that is wanted to enable us to eliminate all the causes that were ever concerned in producing explosions, and perhaps also to put us in a condition to assist, if not in preventing, at least to place some check to the increase of those lamentable occurrences. Bearing in mind what is said in section 28 (page 88) on the subject of surcharged steam, and the peculiar liability there is of boilers exploding exactly at the moment or soon after the valve is opened for starting an engine, there is reason to believe that many explosions may be accounted for by a *coincidence* of two or more distinct causes—one of them always being *the sudden opening of the safety or other valve*—the other either the presence of surcharged steam or an insufficiency of water, or both. For after giving a very attentive consideration of the details of all the explosions that have occurred in Lancashire for a long series of years, I am also of opinion that more than one-half of them have occurred either at that particular time or *after* the *opening* of the safety or other valve, when one might expect it to cause a *diminution* instead of an increase of pressure. The fact is of such frequent occurrence, that I have made several attempts to draw the attention of coroner's juries to the circumstance, but, as may be supposed, as such bodies are ordinarily constituted, with no great success. Every one nearly being now more or less an amateur in steam, the popular idea with juries is that of being appointed to find somebody guilty of overloading or setting the safety valve fast. I am

not going to ascribe any of the above accidents either to surcharged steam or opening of safety valves, because there are
other sufficient causes for them, but the circumstance of explosions so frequently taking place at that precise time is not
the less singular.

Although the American government committee on explosions
did not succeed in proving surcharged steam to be dangerous,
they made the experiment of injecting water upon a red-hot
boiler bottom, when the steam ran up from 1 to 12 atmospheres in one minute, and the boiler exploded with violence.
This is consequently a source of danger which no one ever
disputed. The only question of interest respecting it, is the
manner in which it is brought about.

Section 39.—Explosions from Incrustations.

A boiler bottom getting *red* hot is not likely to take place
gradually while the engine is at work, the water feeding apparatus in order, and the boiler kept clean. But if a boiler is
allowed to become dirty, or covered with indurated or incrusted earthy matter, interposed between the iron and the
water, there is then no difficulty in accounting for such circumstances producing an explosion at any time; and the way
in which it operates towards that end appears to be something
like the following.

It is known that an internal coating of incrustation, or
boiler scale, is liable to crack and separate into large pieces,
which are thrown off from the boiler bottom at some particular
degree of temperature, depending upon the thickness of the
scale and the kind of substance of which it is formed. We
can easily suppose that by a little hard firing, and unduly
heating the boiler, a large portion of scale may be suddenly
detached, uncovering a considerable area at a temperature
something exceeding the *maximum evaporating point*, which
is well known to be considerably under the lowest red heat of
iron. Now, the first effect produced will evidently be a certain
amount of repulsion between the overheated iron and the water,

which may continue for several seconds, and perhaps for a few minutes; this may account for the sudden *decrease* in the supply of steam that has sometimes been observed just before the explosion of a boiler has taken place.   There must also be a gradual diminution of temperature during this short space of time, in that part of the overheated iron which is exposed to the water—creating a contraction of the metal—increasing as the decreasing temperature of the iron approaches the maximum evaporating point, which is at about 350° to 400° Fahrenheit (See sect. 10, p. 29), and causing a corresponding strain on the rivets in the boiler bottom.   The direction of this strain may generally be traced on examining the bottom plates of any old boiler, and will be found to radiate in lines proceeding from the part which has been most acted on by the fire.

The next and concluding step, in case of the metal not being able to withstand the strain caused by its own contraction, will either be a sudden crack or fracture in some direction across this exposed part, or, what most generally happens when an explosion results, the sudden giving way of some bad seam of rivets, which the most nearly coincides or is parallel with the direction of what would otherwise be the true line of fracture.   This may possibly be at some distance from the part which has been overheated, thereby giving the increased effect of great leverage to the pressure acting upon all that portion of the boiler included between the overheated part and the actual line of fracture.   Now the consequence is, not perhaps that this portion is blown out, as would most probably be the case with very *bad* iron, but it will be bent or doubled back, the line of flexure running across the hottest or the weakest part of the iron.   This may help to account for the remarkable way in which we sometimes find exploded boilers twisted and doubled up.   A rupture being thus effected an explosion is inevitable if the hole be sufficiently large.

When an incrusted boiler bottom becomes highly heated,

and the water at the same time too low, it very commonly happens that a large quantity of water is immediately let in, when the consequences are similar to those just described : for the internal coating of scale being suddenly contracted by the cooling effect of the water admitted, it is detached in the same manner as it would have been by the expansion of the iron, and the same effects produced, although perhaps more speedily, as the water admitted will reduce the temperature of the exposed part of the boiler bottom more rapidly to the maximum evaporating point.

Whenever a boiler bottom is seen, or supposed to be, approaching to redness, and that can only happen when the water is *not* boiling, or when the engine is standing, the engine-man should be cautioned against allowing a fresh supply of water to go into the boiler, whether the boiler is short of water or not, until after the engine has been some time at work. My advice to engine-men in such cases is *not to start the engine at all*, but to open the fire doors and stand at a safe distance until all goes cool. I would not have him stop to pull the fires out, and on no account to open the safety valve, as being little less hazardous than starting the engine. If, indeed, he knows the safety valve to be overloaded or made fast, and the steam still continues to rise with the fire doors open, the fires may then be quenched by a jet of water from a hose pipe, or other safe means.

Section 40.—Deposit of Sediment.

All natural waters hold various solid matters in solution or suspension; when in the latter state they admit of being removed by filtration; but no system of filtration, on a scale sufficiently large to supply a moderate-sized steam engine at a light expense, has yet come into practical use. However, it occurred to a gentleman several years ago to hit upon a very simple and effectual substitute. And that was, instead of separating the water from the dirt, before passing it into the boiler, he separates and *collects* the dirt from the water after it is in the boiler by means of a series of vessels, shelves, or trays, placed up and down the boiler, constituting, in fact, so many portions of what collectively might be considered a substitute for a false bottom, upon or into which all the matters held in suspension are deposited. This, in fact, is the whole of the principle of Mr. Anthony Scott's patent of 1827, which has been so frequently re-patented and re-registered since that time, like many bad copies of good pictures, some of them so very bad that the patentee, if he were living, would not know that they were even meant for imitations.

These sediment vessels operate much after the same manner as certain quiet still places do along the banks of rivers, in causing sand or mud to accumulate in them; making so many places of shelter, where any moveable matters being accidentally deposited, they remain free from agitation and not disposed to move out. In a boiler containing *boiling* water, of course the same principle prevails; the steam rising from the boiler bottom—the sole cause of ebullition in all cases—being the agitating agent. In fact the water never boils within the internal vessel or sediment receiver, however violently it may boil externally: and the more violently the water boils the more rapidly the internal vessel collects all loose sediment floating about in the water. Hence Mr. Scott called any internal vessel or apparatus put into a boiler for this purpose, a " sediment collector."

The great merit of Mr. Scott's invention is its peculiar simplicity and cheapness; this was so obvious, that many hundreds of boilers were immediately fitted up with collecting vessels of various kinds with more or less success. Excepting for calcareous incrustations, the process was perfectly successful in keeping a boiler clean. The only difficulty in its practical application was liability to neglect in cleaning out the collectors themselves when they got filled with deposit, and the necessity of emptying the boiler for that purpose.

For the above reasons it appeared desirable to the patentee to have his cleansing apparatus made *self-acting,* that is, to

clean itself out without interruption to the working of the
engine, or letting down the steam ; and he honoured me with
a commission to make such an improvement, which I succeeded in effecting in 1829, when the first complete boiler-
cleansing *machine* was executed and applied to a boiler at the
calico-printing works of Messrs. Thomas Marsland and Son,
in Stockport, who afterwards had fifteen boilers so fitted.
Since the above period they have continued in general use in
Lancashire.

The general form of this apparatus, and mode of fixing it
up within a wagon boiler, is shown in the 1851 edition of
Tredgold on Marine Engines, page, text, 129.   Many hundreds
have been made and adapted to various kinds of boilers, including those of railway locomotives and steamboats.   In the last-
mentioned cases, and in all cases where there is no fire *under*
the boiler bottom, they are, generally speaking, unnecessary,
except for the purpose of preventing priming, which they
most effectually do when that arises from dirty water.   For
this purpose the upper conical-shaped vessel is made with the
narrow collecting apertures adjusted partly above and partly
below the surface of the water.   In this way it is used by
opening the valve at the end of the boiler, and putting the
handle of the agitator in motion for half a minute, by which
the contents of the receiver at the bottom of the boiler are
discharged upwards through the pipe on the right hand.
This operation creates a current, which draws all the *scum*
and *froth* that *cause* the priming from all parts of the water
surface into the collecting vessel and down into the receiver,
whence they are discharged to the outside of the boiler by a
repetition of the process.

By thus *skimming* the dirt from the top of the water, clean
dry steam is supplied to the cylinder of an engine instead of a
mixture of steam and dirty water, causing, in ordinary cases,
such great waste of power by friction on the piston and piston-
rod, and unnecessary consumption of tallow.

Great consumption of lubricating material is always proof

of imperfection in machinery. Instances are not wanting of large stationary engines working for months together without grease of any kind to the piston-rod,* contrasting greatly with the lavish use of that material in marine engines, rendered necessary mainly by the greater liability to prime. The old device of throwing tallow into a steamboat boiler, in order to *prevent* priming is still without a satisfactory theory; unless, as I am inclined to believe, the practice is only beneficial rather in *mitigating* the immediate effect than in the prevention of priming. The only suggestion to account for it that I have met with worth attention, is one by Mr. W. Keld Whytehead, C.E., in an article on the priming of boilers, in the "Artizan Journal" for December, 1848, in which he supposes that, as the tallow requires a very high temperature to vaporise it, it "consequently floats *like a hot plate on the surface of the water,* and tends to separate the particles of water from the steam as they rise."

Fig. 18.

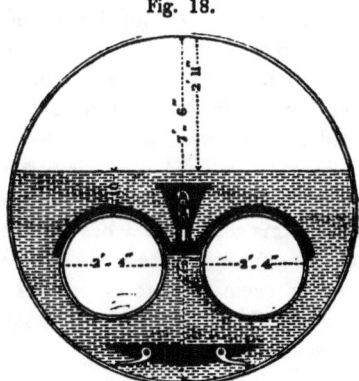

In applying the cleansing machine to boilers containing inside flue tubes, it is necessary for their perfect efficiency

* Should this fact be doubted, I can refer to a case of the kind in London, at Messrs. Esdaile and Margrave's, City Saw Mills, Regent's Canal; their boilers being furnished with my cleansing machines, and the piston-rod stuffing boxes packed with their patent cork fibre.

generally to place two in each boiler, one on each side of the
tube. This is more especially requisite in boilers about 30
feet long, so that one machine may be fixed at the extreme end
of the boiler from the fire, and the other about the middle.
A nearly similar plan is adopted in placing the common sedi-

Fig. 19.

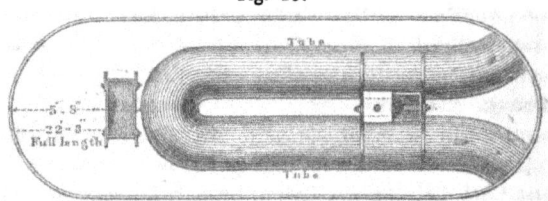

ment collectors as represented in figs. 18 and 19, which are
vertical and horizontal sections (the latter to a scale of half
the size of the former) of the "egg-*ended*"* double-flued
boiler as now made at the principal iron works in Staffordshire;
showing also my last improved modification of Scott's patent
apparatus. The lower collector is of the form originally
recommended by Mr. Scott, and is merely a shallow tray
standing on four legs, about 6 inches from the bottom of the
boiler, calculated only to retain some of the heavier kinds of
deposit. The upper collector is a cast-iron trough suspended
between the flues, and partially or wholly covered with one or
two cast-iron lids; this collector catches and retains the finer
kinds of deposit that are floated near the surface.

Half a dozen of these collectors are well calculated for
placing in a large boiler where it is necessary to use any
material for preventing incrustation.

* This is an excellent form of boiler, but it must not be confounded
with the " egg-*shaped*" or *furnace* boiler more recently brought into use,
a description of which more properly belongs to the subject of Chimneys
and Furnaces.

Section 41.—Calcareous Incrustations.

When the incrustation forming on the inside of boilers consists principally of argillaceous or silicious matters, it is easily prevented by the use of one or the other of the above described apparatus. When, however, any considerable proportion is either *carbonate* or *sulphate of lime,* considerable difficulty is experienced in preventing its formation to an injurious extent. The latter substance more especially, it is well known, has withstood all attempts at complete prevention by chemical means, except such as would be also injurious to the iron. The principal remedial agent that has been found beneficial in any degree to mitigate the effects of this substance, is *crushed potato,* which does not act chemically, but mechanically, the pulp of the potato being supposed to envelop the crystals of the sulphate of lime as they form, and prevent their adhesion to each other.

With respect to the incrustations of *carbonate of lime,* the case is very different. It admits of various methods of preventing its formation by chemical reagents. The means are generally very simple, but like *smoke burning* the question of the best cure for boiler scale has got entirely into the hands of the "chemists *and druggists,*" and opened up a boundless field of quackery and pretension which it would be impossible to characterise properly. I shall therefore only briefly mention a few that I believe to be exceptions to this remark.

The most popular of the patent remedies is that of Dr. Ritterbandt. This plan is to put into the boiler daily a small quantity of muriate of ammonia, or sal-ammoniac, the effect of which is, that any bi-carbonate of lime in solution in the water is decomposed, the muriatic acid of the muriate of ammonia taking the lime and *keeping it in solution,* while the carbonic acid joins the ammonia, forming carbonate of ammonia, which passes off along with the steam. It need not be observed that this remedy can have no effect whatever in preventing the sulphate of lime incrustation.

The theory of the following remedy is something like the reverse of the foregoing, and was, I believe, first proposed by Mr. John Graham, of the firm of T. Hoyle and Sons, and has been several years in use at their celebrated ·calico-print works in Manchester.   It is to put into the boiler daily or weekly a quantity of quick or newly-slaked lime, the effect of which is to convert the *soluble bi-carbonate* into the *insoluble carbonate* of lime, which, instead of being kept in solution as muriate of lime is in Dr. Ritterbandt's remedy, is precipitated and collected without any trouble by sediment collectors, or collected and discharged from the boiler by the cleansing machine.

This putting in of lime to take out lime is a nice application of Dr. Clarke's simple and efficacious method for purifying water on a large scale, now so well known and generally approved by water-works' companies.

In Lancashire, where generally a great portion of the boiler scale is sulphate of lime, it has long been a practice to use ox-feet, or any animal substance convertible into jelly by boiling, with good effect.   But they are liable to promote priming, and, like potatoes, they require frequent renewal.

One English patent, now expired, specifies the use of all kinds of vegetable matter or extract without exception, preferring that which gives out the greatest quantity of *colouring matter*, as logwood, bark, or tan.   Also turf, peat, manure, leaves, saw-dust, and charcoal.   Other patentees recommend urine, glue, blood, dung, and night-soil.   Also sugar, starch, treacle, flour, malt, and the bottoms or settlings of beer barrels. Most of the above articles may be used with advantage where there is not much of the sulphate, but they all act mechanically.

Tan and salt are the principal ingredients in some of the best of the foreign patents, which generally also contain some corrosive materials that are difficult to particularise and hazard-ous to use.

APPENDIX.

THE following very able report, made by the Association in Manchester, for preventing steam boiler explosions and for effecting economy in the raising and use of steam, is too valuable to be withheld from this work, and which is appended for the instruction of our readers:—

" The short period which has elapsed since the association commenced operations, and the time absorbed in collecting and registering the necessary details incidental to every first inspection, has scarcely permitted the full regulation of all the benefits which the association is calculated to confer; but it is satisfactory to state, that no explosion or accident has occurred in any boiler under the supervision of the association. The chief inspector has, however, reported several cases of imperfection tending to accidents, and, in particular, has found many flues so constructed as to transmit heat directly to the steam in the boiler, not only when the water is deficient but when at its daily working level, thus surcharging the steam with heat, and endowing it with one essential element of explosive power, which may be instantly developed by the admixture of water, by agitation or otherwise. The fact that steam, in contact with water in a quiescent state, may be heated to 500° or upwards, without any corresponding effect on the steam gauge, or proportionate increase of pressure, appears to be established on good authority. But the precise condition under which the surplus heat thus accumulated in the steam may combine with water to produce explosion is not fully known; and it is recommended that an investigation of this important point by experiment, as a proper subject for this association to determine. In like manner the economical effect, if any, obtained by heating steam in its passage from the boiler to the steam cylinder, for the purpose of gaining pressure by its expansion demands investigation; as also the still more important question of the strength of plate-iron tubes or internal flues to resist external pressure, for estimating which no data has yet been made public. In the important branch of effecting economy in the raising and use of steam some progress has been made in collecting facts bearing on the question, and sufficient have been obtained to prove a great waste of fuel in many establishments. It has been ascertained that the consumption of coal for indicated horse power per hour ranges in different cases from three to twelve pounds, and after making every allowance for differences in the quality of coal, and for the employment of part of the steam in heating and other purposes, there remains a vast field for pecuniary saving. It is also found that the waste lies not only in the faulty construction and ill-adapted proportions of furnaces and boilers, but also in the mode of applying and using the steam when raised. The clouds of dense smoke, indicative of imperfect combustion, testify to the defects of too many furnaces, and indicator diagrams attest in like manner the waste of steam, showing in some cases, in which high and low pressure cylinders are worked together, a loss of 10℔ or upwards of pressure between the exhaust side of the one cylinder and the steam side of the other. We may look forward with confidence to important improvements in these particulars, by the daily experience in inspection now acquiring and in the supervision provided."

Mr. Whitworth read the following report, made to the committee by Mr. Longridge the chief inspector:—

In laying before you a report of our proceedings from the 1st April to the 17th of the present month, I shall first state the chief subjects which have been proposed for our investigation; and in the next place show how far we have been enabled to prosecute these inquiries, and what conclusions may be fairly deduced from them. Our attention has been directed —1. To an examination of all boilers placed under our inspection, with a view to ascertain as correctly as possible their actual condition, and whether they were adapted to their ordinary working pressure; also, whether they were provided with the requisite mountings, and if the same were kept in good working order.—2. In those cases of explosions which have taken place in the neighbourhood, to an investigation of the peculiar circumstances connected therewith, in order, if possible, to ascertain the real cause, and the best means of preventing the recurrence of such accidents.—3. To ascertain by comparison the most advantageous construction, dimensions, and working of boilers in regard to safety, economy of fuel, and durability.—4. To ascertain the most economical system of employing steam as a motive power.

The number and description of boilers at present under our inspection are as follows, viz.:—

Description.	Pressure per square inch.						
	15℔ or under.	16℔ to 30.	31℔ to 45.	46℔ to 60.	61℔ to 75.	76℔ to 80.	Number.
Cylindrical, with internal flues	83	274	144	60	28	—	589
Cylindrical, without do.	10	12	10	6	7	—	45
Galloway's patent boilers	2	35	48	12	—	—	97
Multitubular . . „	12	9	31	32	2	1	87
Butterley . . „	43	20	1	—	—	—	64
Wagon . . . „	38	—	—	—	—	—	38
Total	188	350	234	110	37	1	920

Of the above, 81, or nearly nine per cent., have been found to be in a dangerous state, from the following causes, viz. ;—

Construction or strength not adapted to the working pressure . 24
Defects in the plates or angle iron 9
Defects in the boiler mountings 26
Injury sustained from deficiency of water 19
Ditto „ deposit of scale 3
                                                          ———
                            Total 81

In addition to the above 19, rendered dangerous from deficiency of water, there are 14 others which have been injured to a less extent from the same cause. This is evidently the most frequent cause of explosion, as will be explained hereafter; and as it is important to provide such means of prevention as will be effective in cases of negligence on the part of the fireman, I would suggest, first, the general adoption of open stand pipes, where applicable, or safety valves, in connection with a float, to allow the escape of steam, whenever the water falls below the fixed limit.—2. The use of fusible metal plugs, fixed on the top of the flues above the fire. These should stand sufficiently high to melt before any part of the flues could be uncovered with water. The usual practice of inserting a lead rivet or plug in one of the plates is worse than useless, inasmuch as owing to the inclination usually given in setting boilers, a considerable portion of the flue must be exposed, and may even become red hot, before such lead plug can be melted; under which circumstances an explosion is the probable consequence.

EXPLOSIONS : although there have been no cases of explosion in boilers under our immediate inspection, we have had opportunities of examining the following, viz. :—Cylindrical boiler, with two internal flues at Messrs. G. and G. Pilkington's, Crawshaw Booth, exploded 25th April: cylindrical boiler, without internal flue, at Mr. Ralph Wood's, Salford, 8th June; upright cylindrical boiler, with two internal flues, at Messrs. Beyer and Peacock's, Gorton, 14th June; wagon boiler, at Mr. Samuel Rothwell's, Elton, near Bury, 25th June; cylindrical boiler, with one internal flue, at Messrs. Watson and Allsup's, Preston, 30th July; cylindrical boiler, with two internal flues, at Messrs. William Parker and Co.'s, Sheffield, 11th August; multitubular boiler, at Messrs. Shortridge, Howell, and Jessop's, Hartford-Street Works, Sheffield, 18th August; cylindrical boiler, with one internal flue, at the Lancashire and Yorkshire Railway Works, Miles Platting, 13th October. With the exception of the boiler in Salford, which was not, in fact, provided with any safety-valve; that at Messrs. Parker and Co.'s, in Sheffield, which exploded from excessive pressure; and the old wagon boiler, at Mr. Samuel Rothwell's, near Bury, these explosions may all be traced to the same cause viz.: the production of the surcharged steam, by the transmission of heat, above the surface of the water. Although the possibility of surcharging steam, while in contact with water, is still disputed by many engineers in this country, this question was satisfactorily solved by a committee of the Franklin Institute, in America, above twenty years ago. In the report of this committee, it is stated that "the temperature was carried to 533 degrees Faht., when the pressure, shown by the gauge, was 6·82 atmospheres: while saturated steam, at that temperature, would have had a pressure of more than 60 atmospheres;" and further, "these experiments, which lasted more than two hours, show that the surcharged steam remained in contact with water without acquiring from it the water necessary to convert it into saturated steam, but retaining its surcharged state." Several instances which have come under our own observation might be adduced in confirmation of the experiments of these gentlemen, but I shall only mention one, which lately occurred, as sufficient for our present purpose. The boiler referred to contained two internal furnaces, uniting in one flue, and had been filled with water to the usual height by a pipe leading from a reservoir. The end of this pipe was about 9 inches below the top of the furnaces. About two hours after lighting the fires, the steam (being at 8℔, as indicated by the gauge) was turned into the mill for the purpose of warming it. Shortly after this the attendant observed that the water had disappeared from the gauge glass, and was forced back into the reservoir, the valve on the feed pipe not having been entirely closed. At this time the upper part of the furnaces, above the surface of the water, had become red hot, and the temperature of the steam was such that a block of wood, resting on the top of the boiler, was converted into black charcoal, and yet the pressure never exceeded 8℔. The communication with the reservoir having been closed, the fire doors opened, and the damper shut, the boiler was allowed gradually to cool; and although the tops of the furnaces were depressed, no explosion took place. From this it is evident that steam may be raised to a high temperature, while in contact with water, and yet remain at a low pressure. And this condition can only arise from a deficiency of water in such steam; we may reasonably infer, that if this could by any means be supplied, we should have an almost instantaneous increase of density and pressure proportionate to the degree of saturation. This will fully account for the difference in intensity of many explosions, and why these should so frequently occur immediately after starting the engine, admitting water into the boiler, or lifting the safety valve, all of which tend to produce agitation of the water, and to promote its diffusion amongst the steam. Although this theory of boiler explosions, which was advanced by the late Mr. Perkins many years ago, has not hitherto been generally admitted, certainly the facts which have come under my own observation seem fully to confirm its accuracy. The two last subjects which have been proposed for investigation require a greater amount of data than we have as yet been able to obtain; and although we have had occasion to remark great errors in many of the present modes of employing high-pressure steam, I should not now be justified in expressing a decided opinion as to which system is positively the best, nor which is the best construction of boiler. In connection with the working of boilers, the subject of smoke-prevention has not been overlooked. Our experience on this subject tends to the conclusion, that without much difficulty or expense the smoke nuisance may be greatly abated, in almost every description of boiler, and that this will be accompanied by a saving of fuel, provided such attention as might reasonably be expected be given by the fireman. During the ensuing year these subjects shall have the attention which they require, and I trust, ere the next general meeting of the association, we shall be enabled to arrive at a satisfactory solution of these important questions.—I am, gentlemen, yours respectfully,

ROBT. B. LONGRIDGE, Chief Inspector.

In moving " That the report of the committee be adopted, and printed for circulation," the Chairman said that the objects of the association were to secure greater safety in the working of boilers, and economy in raising and using steam. This was proposed to be done by an intelligent and regular supervision, to be undertaken by inspectors appointed and paid by the association; that inspection not being intended in the smallest degree to diminish the care or responsibility of the members of firms or their servants, but being superadded thereto : in fact, to form, as it were, a guarantee that whatever was dangerous or wasteful in connection with the steam department, should be brought to the notice of members at the earliest possible period, with a view to remedies being applied. It was also intended, that the information and facts obtained by the inspectors in their extensive rounds should be so classified and concentrated, at the head offices, as to be accessible to every member; so that information might be obtained upon any point of practical working upon which difficulty might be experienced. It was evident that the success would very much depend upon the qualifications of the chief inspector—that if an inspector should be selected who, from want of knowledge, industry, or discretion, did not secure the confidence of members, that amount of success would not be obtained without which the association could not be effective for good. It appeared also, that in the inspector, there would be required something more than mere practical every-day knowledge. The use of high-pressure steam was being rapidly extended ; the whole question of the best means of raising and applying steam was in a transition state; and questions were

constantly being raised, with which a man possessing only every-day practical knowledge could scarcely be expected to deal. It was essential to
the permanent success of the association, that the person selected as chief
inspector should be one who had enjoyed the advantages of a liberal edu-
cation—who had acquired sufficient scientific knowledge upon the questions involved in the management of steam to be able, theoretically, to
take up difficulties or suggestions that might arise—who possessed sufficient mathematical knowledge to avail himself of the various published
formulas and data, and to work them out in connection with information
acquired in the discharge of his duties—and who should also be a man of
considerable activity, quick perception, and apt to learn by experience.
Mr. Longridge gained experience from his duties, and would be able to
guide them through the various difficulties that might arise, a good commencement of what would prove a very valuable feature of the association.
There were other means of extending knowledge as to steam, and generally aiding members on that subject, that had not yet been entered upon.
It was intended that, at an early period, inventors should be invited to
send in models or prospectuses of inventions; and that the association
should thus, by having a depository for such things, be able to submit to
members the various instruments and mechanical contrivances which in-
genuity had devised for the purpose of being applied to steam boilers,
furnaces, or engines. Inventors should know that the association had
made arrangements for receiving models, explanatory drawings, and
specifications of patents, or anything in connection with the subjects
mentioned, which it might be wished to bring before the public;
and in a short time, the association would have a gradually increasing
and most important collection, for the information of members. He
would now draw attention to what was conceived to be still wanting, in
order completely to carry out the expectations of the originators of the
association. If it were asked what were the inducements to become
members, he would reply that the pecuniary advantages to be derived—
quite independent of considerations of safety to life and limb—far exceeded any charge proposed to be made, or that it could ever be necessary to make. It was believed that considerable saving might be effected
through improved arrangements for raising and applying steam. The
report stated that, in connection with the boilers inspected, the consumption of coal per horse power varied from 12℔ to 3℔. Special circumstances might account for large consumption in some cases; but, making
every allowance, there was a vast field for saving in that department
alone. In another direction, there was the saving to be effected, where
high-pressure and condensing cylinders were used together; the fact
stated by Mr. Longridge on that subject being very conclusive as to the

loss from defective arrangements. He believed it probable that the members of the association alone paid upwards of £200,000 a year for coal, and that a saving of from £10,000 to £20,000 might be effected in a very short time by the application of principles already known. There was by no means a wasteful consumption of coals in his own concern previously to 1838; but in three years from that time, a saving of 1,100 tons was effected, as a result of his reading Mr. Charles Wye Williams's treatise on combustion.* The consumption had been 20 cwt. an hour, upon the average of three years; but after reading Mr. Williams's work, he caused a certain quantity of atmospheric air to be admitted into the furnaces, instead of having coals constantly shovelled in; and upon the average of the next three years, the consumption was only 17 cwt. an hour. There were thousands of firms in whose establishments there was no external supervision, but in which there might be found the same causes of danger that had been detected, and the same wasteful consumption of coal that had been pointed out, in the establishments of members of the association. The following gentlemen were appointed as the committee:—Messrs. James M'Connel, George Peel, Richard Birley, and James Dugdale, Manchester; James Platt, Oldham; Joshua Radcliffe, Rochdale; William Wanklyn, jun. Albion Mills, near Bury; Edward Birley, Preston; Alfred Neild, Manchester; Richard Peacock, Gorton. Messrs. T. Cooke and R. Johnson were appointed auditors.—Mr. Fairbairn trusted that the result of the experiments would be at least to solve the questions as to explosion from collapse, or from external pressure upon flues. He also hoped that valuable results might be obtained with respect to surcharged steam; and as to whether or not it was advantageous to heat steam after it had left the boiler. He thought the association promised well for the next year; and he trusted that the association would determine that, out of the funds of the association, experiments should be made, at least so far as to elucidate those points in which the members were more directly interested, as bearing upon the practicable operations of their businesses. The association was the first of its kind, he was very anxious that the experiments should be conducted in connection with the body. He also particularly desired the aid of their intelligent chief inspector, who would, he was sure, take an active part in conducting or assisting in such experiments.

SMOKE PREVENTION.

The Smoke Nuisance has assumed a degree of public importance it never before possessed, the Legislature having passed an act embodying a conviction that the prevention of smoke is possible, and levying fines in all

* 2d edition 8vo., 1854, Weale.

cases where the nuisance is continued. It was, in fact, distinctly proved, previous to the passing of the " Smoke Prevention Bill," that furnaces could be so constructed as to consume their dense noxious vapour. Mr. C. Wye Williams's Smoke Prevention Apparatus, consists simply of a practical application of the sound chemical principles so long contended, in the face of popular prejudice and no small share of imaginary self-interest. Mr. Williams's apparatus is simplicity itself, merely because the principle it involves is sound.

Hitherto, the practice has been to exclude as much air and heap on as much coal as possible, thus preventing a sufficient quantity of the common atmosphere getting into the furnace to secure complete combustion. The consequence was the gasses were too slowly evolved, and passed away into the chimney in dense black columns, to poison the air of the surrounding neighbourhood. This is prevented now, however, by a moderate enlargement of the fire-beds and flues, and the introduction of air to the surface of the fire through perforated doors, and plates placed between them and the fire. The furnace itself being constructed to admit the quantity of air required for perfect combustion, the perforated plates secure such a mechanical division and distribution of the common atmosphere as to ensure its becoming instantaneously heated, and promoting, instead of retarding, as a column of cold air does, in a great degree the combustion sought. The whole mystery of all the smoke preventing apparatuses now in public favour lies in this very simple secret.

We will proceed to give an analysis of experiments which have been repeatedly made with Mr. Williams's apparatus. With the air wholly excluded, 251 ℔s. of coal were consumed per hour, 1,239 ℔s. of water were evaporated, the Pyrometrical heat being 381°; whereas, with the air admitted, the consumption of coal was only 236 ℔s. per hour, the water evaporated 1662 ℔s., and the temperature, as indicated by the Pyrometer, 901°; thus showing an immense saving of fuel and an extraordinary increase in the quantity of water evaporated, as indicated by the Pyrometic action.

These phenomena are of immense commercial value. So much so, that those who employ steam should adopt the new furnace, being vastly to their advantage, more particularly as the cost attending its introduction, in places were the largest amount of steam power is employed, would not exceed a very insignificant sum, as compared with the saving of fuel. On looking into the flues of Mr. Williams's apparatus, the whole secret discloses itself to the observer. The air is evenly distributed over the whole face of the fire, and becoming, as we before observed, instantaneously heated, by the time it reaches a certain point in the flue, complete combustion has taken place; and there being no smoke, the flues can be seen through with the naked eye. But the moment the air is shut off, the furnace and all its passages instantly become dark, and filled with black smoke, thereby proving that air, when judiciously introduced, is the most indispensable element in producing the chemical action required.

The universal applicability of this principle cannot be disputed; and its use in private dwellings, which are, after all, the great vitiators of the atmosphere in all our large towns, is only a question of time. Still, there are many who object to introducing anything new; but they should bear in mind, that Mr. Williams's furnaces have been in use in the City of Dublin Steam Packet Company's boats for some years, and that the law requires them to adopt means not for the partial, but the entire removal of the smoke nuisance, no matter whether it originates with the malconstruction of the furnaces now in use, or a dislike to encourage anything like an innovation of what has hitherto formed a very important part of the business of our motive-power engineers.

Considerable opposition has been raised to Mr. Williams's furnace, on the supposition that the admission of air on his principle is tantamount to throwing open the doors of the furnace, from which, a loss of fully 50 per cent. of the steam-generating power of the fire ensues. The analysis above given proves, however, that this is quite a mistake. Where the furnaces and flues are constructed on the strict scientific principle laid down by Mr. Williams, a far greater amount of steam is generated, and with considerably less fire, than with the ordinary furnaces, and it is only where the fire beds and flues are too small that Mr. Williams's apparatus cannot be employed with success. On this subject, however, we cannot do better than quote a short passage from a speech recently made by the eminent engineer, Mr. Houldsworth, at the annual meeting of the " Association for the Prevention of Steam-boiler Explosions :"—" There was by no means a wasteful consumption of coals in his own concern previously to 1838, but in three years from that time a saving of 1,100 tons was effected, as a result of his reading Mr. Charles Wye Williams's treatise on combustion. The consumption had been twenty cwt. an hour upon the average of three years; but after reading Mr. Williams's work he caused a certain quantity of atmospheric air to be admitted into the furnaces, instead of having coals constantly shovelled in; and upon the average of the next three years the consumption was only seventeen cwt. an hour."—*Northern Daily Times, Dec.* 11, 1855.

Again, the smoke nuisance having given much cause of complaint, —the recent Act has had but partial success,—we will conclude by the following trite remarks made by Dr. Ure in his Dictionary of Arts, &c. :—" Among the fifty several inventions which have been

H

patented for effecting this purpose, with regard to steam-boiler and
other large furnaces, very few are sufficiently economical or effective. The
first person who investigated this subject in a truly philosophical manner
was Mr. Charles Wye Williams, managing director of the Dublin and Liverpool Steam Navigation Company, and he also has had the merit of con-
structing many furnaces, both for marine and land steam-engines, which
thoroughly prevent the production of smoke, with increased energy of
combustion, and a more or less considerable saving of fuel, according to
the care of the stoker.   The specific invention, for which he obtained a
patent in 1839, consists in the introduction of a proper quantity of
atmospheric air to the bridges and flame-beds of the furnaces through *a
great number of small orifices*, connected with a common pipe or canal,
whose area can be increased or diminished, according as the circumstances
of complete combustion may require, by means of an external valve. The
operation of air thus entering *in small jets* into the half-burned hydro-
carburetted gases *over the fires, and in the first flue*, is their perfect
oxygenation—the development of all the heat which that can produce,
and the entire prevention of smoke.   One of the many ingenious methods

Fig. 20.

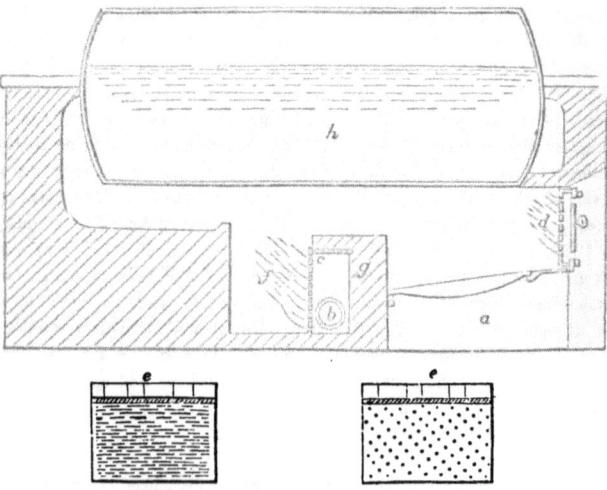

in which Mr. Williams has carried out the principles of what he justly
calls his argand furnace, is represented at fig. 20, where *a* is the ash-pit
of a steam-boiler furnace ; *b* is the mouth of a tube which admits the

external air into the chamber, or iron box of distribution *c*, placed immediately beyond the fire-bridge *g*, and before the diffusion, or mixing chamber *f*. The front of the box is *perforated either with round or oblong orifices*, as shown in the two small figures *e e* beneath ; *d* is the fire-door, which may have its fire-brick lining also perforated. . In some cases the fire-door projects in front, and it, as well as the sides and arched top of the fire-place, are constructed of perforated fire-tiles, enclosed in common brickwork, with an intermediate space, into which the air may be admitted in regulated quantity through a moveable valve in the door. I have seen a fire-place of this latter construction performing admirably, without smoke, with an economy of one-seventh of the coals formerly consumed in producing a like amount of steam from an ordinary furnace."

January 25th, 1856.   J. W.